CONTEMPORARY CITY ECOLOGY

PERSPECTIVES IN URBAN GEOGRAPHY
VOLUME SIX

Concept's International Series in Geography

No. 1 Perspectives in Agricultural Geography (In Five Volumes) Edited by NOOR MOHAMMAD

No. 2 Perspectives in Geomorphology (In Four Volumes) Edited by H. S. SHARMA

No. 3 Perspectives in Urban Geography Edited by C. S. YADAV

1. New Directions in Urban Geography
2. Comparative Urban Research
3. Comparative Urbanization: City Growth and Change
4.A Models in Urban Geography (Theoretical)
4.B Models in Urban Geography (Mathematical)
5. Urban Research Methods: Central Place, Hierarchical and City Size Models
6. Contemporary City Ecology
7. Slums, Urban Decline and Revitalization
8. Contemporary Urban Issues
9. Rural-Urban Fringe
10. Morphology of Towns
11. Cities and Housing
12. Perceptual and Cognitive Image of the City
13. Urban Economics
14. City Planning: Problems and Prospects
15. City Planning: Administration and Participation
16.A City Planning and Policies: Reorientation of Policy Norms
16.B City Planning and Policies: Cross-cultural Urban Policies

Concept's International Series in Geography No.3

PERSPECTIVES IN URBAN GEOGRAPHY

VOLUME SIX

CONTEMPORARY CITY ECOLOGY

Edited by
C.S. YADAV

CONCEPT PUBLISHING COMPANY, NEW DELHI-110015

First Published 1987

Published by

Naurang Rai
Concept Publishing Company
H-13, Bali Nagar
NEW DELHI-110015 (India)

Printed in India by

New Gian Offset Printers
495, DDA Complex, Shahzada Bagh
NEW DELHI-110035

Preface

NATURE OF URBAN GEOGRAPHY

THERE are several studies on urban geography which reveal that, as a major sub-discipline, it has a long tail but a short body. However, the spectrum is so vast and broad that researchers are able to make general statements in defence of its conception, philosophy, nature and orientation. Urban geography today encompasses and interfaces with various disciplines which are interested in urban studies. As an introduction to this series, our task here is to make an attempt to briefly review the development of geographical interest in various aspects of cities.

There are several reviews of the early development of urban geography which have been dealt with by Berry and Harton (1970)[1] and Carter (1974).[2] Berry and Harton in their book, **Geographical Perspective on Urban System,** have made an attempt to introduce readers to its present-day status. They have clearly resolved that "the formative years of the social science in the late nineteenth century and early twentieth century were also the years in which urban studies first developed,"[3] thus providing the context for geography's emerging interest in cities. However, the emergence of urban studies dates back to the writings of Greek scholars, but as a sub-discipline, it has reached its present-day status only in the past 30 years.

In their historical perspective the works on urban geography show that the pre-20th century studies primarily concerned themselves with themes of location, size and shape of the cities. The initial findings were strongly subjective, descriptive and dependent more on observation such as the works of Hassert (1907)[4] and Blanchard (1911).[5] In the succeeding years the conceptual framework of site and situation was criticized by Aurousseau (1924)[6] and Crowe (1938)[7], their conception being that cities were not inanimate objects in landscape, but also organic elements which involved people and their movements. The morphologist, later in 1960, truly brought the indigenous line of evolution in the sphere of urban geography and studies on the build-up fabric of cities (Conzone, 1960).[8] Smails (1955)[9] constituted the prime base of urban geography, which remained articulated without any major conceptual change till early 1960s. The studies on the morphological aspects of the urban system were influenced by external forces, consequently the methodological frameworks got impetus within the discipline. At this juncture the evolution of concepts was not based on environment but took shelter under the umbrella of economics of location, and incorporated analysis of land values and rents, and the concept of nodality and accessibility. These concepts were derived from the economic theories of Cooley (1894)[10]. Weber (1899),[11] and Hurd (1903).[12]

The Chicago School of Urban Ecology hastened the evolution of urban geography. In his monumental work Park (1925)[13] developed the idea of order and analysis of towns. Further, a powerful thrust and much of the rationale was provided to the studies of urban geography by the Central Place Theory of Christallar (1933).[14]

The impact of the statistical method was powerful and it brought rapid and enormous changes in the field of urban geography. It also brought new insights into the development and application of urban geography towards increased quantification. Brian J. Berry, a pioneer in the field, analysed the spatial order, size and location of towns and cities. There was widespread use of innovative techniques to explore the nature of urban problems, hypotheses were tested, new theories propounded and old theories remodelled. The statistical methods were put to a variety of uses. Smith (1965)[15] evaluated the classification of settlements; Berry and Garrison (1956)[16] examined the utility of the rank-size rule for urban populations. The models of Park and Burgess (1925),[17] Harris and Ullman (1945)[18] and Homar Hoyt (1939)[19] were tested and re-examined by various geographers.

A new impetus to urban geography came from social area analysis which was initially propounded by Shevky and Bell (1955)[20] and later on this provided a basis for factorial ecology. The collaboration of this stream in urban geography was offered by Berry (1971)[21], Herbert (1972)[22] and Johnston (1971).[23] As a consequence of these thrusts in geography after 1960, the techniques of investigation were sharpened and this has provided the basis for a scientific explanation of cities. Attempts were made to introduce new theories and frame laws to make the explanation of events more rational and logical. Sophisticated models were propounded as urban geography entered a new era of rationalizing the subject matter of urban studies on the basis of new philosophies, new concepts, new theories, new methodologies and applications.

However, the status of contemporary urban geography has been elevated only recently by the behavioural approach. The studies of perception and cognition which have a long tradition in physiology were first introduced into the field of geography by Lynch (1960)[24] Dowson and Stea (1973)[25], Gould and White (1974).[26] At present there is a sudden spurt in the studies on the subject with a new paradigm. But the full impact of behavioural approaches upon urban geography has yet to be realized.

The aim of this series is to seek reorientation of the discipline strengthened by new philosophies, methodologies, subject matter or application. The series has been arranged in such a way that all contemporary viewpoints are covered comprehensively. Hopefully, this series will inspire researchers to appreciate the work already done by geographers in studying cities. Geography by nature seems to be a synthesizing field of inquiry. As such we have made an endeavour to combine some of the otherwise disparate facts garnered by other disciplines in such a manner that we can gain a better understanding of the urban system. The study of urban geography is essential if we are to analyse the human consequences of the settlements in which we live. It is useful to planners, decision makers in government and corporations and also to each one of us as citizens. Finally, it gives us a perspective on what may be happening to our cities and to the nature.

To achieve the above goal only those contributions with originality and contrasting viewpoints were selected for inclusion in this series. In doing so the editor does not wish to compete with the journals in the field in which innovative research and methodological aspects are presented. But he does wish to convey, and convey with conviction, that significant researches are being undertaken in the different branches of urban geography and in other allied disciplines. The present series is an attempt to provide a selective reappraisal and rigorous examination of the assumptions and the urge to disseminate new knowledge created by the mutual interaction. Emphasis is also placed on the conceptualization and theorization of the subject matter so that general laws may emerge. A conscious effort has been made to organize the series in such a way that it reflects the philosophical approach parallel to that of the behavioural school. Finally, a vigorous attempt has been made to demonstrate throughout the series how geographers are basically involved in solving the urban problems.

C. S. YADAV

REFERENCES

1. Berry and Harton, **Geographic Perspectives on Urban Systems: with Integrated Readings**, New Jersey, Prentice Hall, Inc, Englewood Cliffs, 1970.
2. Carter, **The Study of Urban Geography**, Edward Arnold, London, 1974.
3. Berry and Harton, 1970.
4. Hassert, K., **Die Stadte Geographisch Betrachect**, Leipzig, 1907.
5. Blanchard, R., **Grenoble: Etude de Geographic Urbanine**, Paris, 1911.
6. Aurousseau, M., "Recent Contributions to Geography: A Review," **Geographical Review**, 1924, 14, p.444.
7. Crowe, P.R., "On Progress in Geography," **Scottish Geographical Magazine**, 1938, 54,1-19.
8. Conzone, M.R.G., "The Plan Analysis of an English City Centre," in Norkorg (Ed), **Proceedings of I.G.U. Symposium on Urban Geography**, C.W.K. Gleerup, Lund, 1962.
9. Smails, A.E., "The Urban Mess of England and Wales," **Transactions and Papers**, Institute of British Geographers, 1946, 11, 85, p.101.
10. Cooley, C.H., "The Theory of Transportation," **Publications of the American Economic Association**, 1894, 9, p.5-7.
11. Weber, A.F., **The Growth of Cities in the Nineteenth Century: A Study in Statistics**, New York, 1899.
12. Hurd, R., **Principles of City Land Values: New York Record and Guide**, 1903, p.19-21.
13. Park, R.E., "Suggestions for the Investigation of Human Behaviour in an Urban Environment," in R.E. Park, E.W. Burgess, R.D. Mackenzie (eds), **The City**, University of Chicago Press, 1925.
14. Christallar, W., **Central Place in Southern Germany**, Gustar Fischer, Teno, 1933.
15. Smith, R.M.T., "Method and Purpose in Functional Town Classification," **Annals of the Association of American Geographers**, 1965, 55, pp.539-548.
16. Garrison, W.L., "Applicability of Statistical Inference to Geographical Research," **Geographical Review**, 1956, 46, pp. 427, 429.

17. Burgess, E.W., "The Growth of City," in R.E. Park, E.W. Burgess and R.D. Mackenzie (eds), **The City**, University of Chicago Press, 1925.
18. Ullman, E.L., "The Nature of Cities," **Annals, American Academy of Political and Social Science**, 1945, 242, 7-17.
19. Hoyt, H., **The Structure and Growth of Residential Neighbourhoods in American Cities**, Washington, Federal Housing Administration, 1939.
20. Shevky, E. and Bell, W., **Social Area Analysis**, Stanford University Press, Stanford, 1955.
21. Berry, B.J.L. (ed), "Comparative Factorial Ecology," **Economic Geographic (Supplement)**, 1971,47.
22. Herbert, D.T., **Urban Geography: A Social Perspective**, David and Charles, Newton Abbot, 1972.
23. Johnston, **Urban Residential Pattern**, Bell, London, 1971.
24. Lynch, **Image of the City**, Massachusetts, Cambridge, 1960.
25. Dowson and Stea, "Image and Environment", Chicago and London, 1973.
26. Gould and White, R., **Mental Maps**, London, 1974.

Acknowledgement

We Thank Editors Gerd-Michael Hellstern, Frithjof Spreer Hellmut Wollmann, **Heft 2, Vol. II and Vol. III** for their permission to use material from "Applied Urban Research", Proceedings of the European Meeting on Applied Urban Research, Essen, Oct. 2-4, 1981.

Contents

List of Contributors

Erling Berge: Institute of Applied Social Research, Oslo-2, Norway

Dennis Conway: Assistant Professor, Indiana University, Bloomington, Indiana

Pierre deVise: School of Urban Sciences, University of Illinois, Chicago circle, Box 4348, Chicago, IL 60680 USA

Ashok K. Dutt: Professor, Geography and Urban Studies, University of Akron, Akron, Ohio

Gerardine D'Sa: Teaching Assistant, Department of Geography, University of Akron, Akron, Ohio.

Peter M. Foggin: Department of Geography, Montreal University

Derek R. Hall: Sunderland Polytechnic, Sunderland, U.K.

Jonathan Kelley: Department of Sociology, Research School of Social Sciences, Australian National University, Canberra, A.C.T. 2600

Elisabeth Lichtenberger: Institut for Geographie der Universitat Wien, Prases des Studienzweiges Raumforsching und Raumordnung, Vienna

Gill C. Lim: Woodrow Wilson School of Public and International Affairs Princeton University, Princeton, N.J. 08544, U.S.A

E van Loggerenberg: Senior Researcher, Institute for Urban Studies, Rand Afrikans University, Johannesberg

Ian McAllister: Department of Sociology, Research School of Social Sciences, Australian National University, Canberra, A.C.T. 2600

Jean-Pierre Thouez: Department of Geography, Montreal University

Prof. J.N. Scheepers: Professor in Geography and Chairman, Department of Geography, Rand Afrikans University, Johannesberg

Christina R. Victor: Research Officer, St. David's Hospital, Cowbridge Road, Cardiff, CF19TZ

Gerald Walker: Associate Professor, Department of Geography and Social Science, York University, Toronto, Canada

C.S. Yadav: Department of Geography, Bhagat Singh College, Delhi University, New Delhi

Gabriele Zanetto: University of Vinice, Italy

ONE

C.S. Yadav

Introduction

Most of the urban geographers have agreed and oriented their discussions to current urban problems known as ecological problems. Numerous academicians have adopted an interdisciplinary approach to urban ecological issues with a view to spanning traditional disciplinary boundaries and gaining a comprehensive understanding of the contemporary patterns, processes, issues and changes in urban environment. In essence, urban ecology sheds light on the total environment of the city in which it studies the elements of its structure, identifies the patterns which they formed and makes an attempt to understand the interrelationships which existed. The basic fundamental principle of urban ecology is that everything is connected with everything else Barry Commoner (1971).[1] Thus the apparent objective of the ecological approach is to describe and analyse the system of interdependence among different elements in a common setting Gist and Fava.[2] As it is self-explanatory in an ecological system, every action has its consequence. In principle, while human ecology predominantly analyses the interrelationships among men in their spatial setting, the urban ecologist more particularly is concerned with these interrelationships as they occur in the city. As a matter of fact, cities offer various dimensions to an ecologist, which include the study of external expressions of ecological interrelationships as reflected in the distribution of cities, their internal structure and composition, changing skyline, alternating land uses and intimate intermingling and patterning of different urban landscapes.

However, despite these apparently laudable objectives, the impact of ecological approach on the development of urban geography is relatively brief and it has a chequered history. In essence, the dominance of ecological approach is relatively short. Nevertheless, the period between 1938 and 1945 witnessed a sharp criticism of both its theoretical framework and its empirical generalizations. The period of the 1950s was the dark period for the ecological approach. However, the appearance of Hawley's Human Ecology, A Theory of Community Structure[3] gave a new lease of life to the ecological approach and virtually initiated its present revival. Eventually, this was the era in which there was a conceptual transformation in urban geography which witnessed parallel streams of research Brian J. Berry and J. D. Kasarda (1977).[1] As a consequence there was a healthy intellectual dialogue among the disciplines in the context of urban studies. Berry and Kasarda rightly remarked that, "Today the ecological approach is generally recognised as an important mode of inquiry, but, because of its history of shifting, substantive objectives and the diversity of its methodological techniques, confusion as well as debate continue to surround its role not only in sociology but in geography as well."

DEVELOPMENT OF ECOLOGICAL APPROACH

The term ecological approach has been overused and generally misused. In its original conception, the term was primarily applied to the physical sciences of botany and zoology, whose aim was to trace the relationship of plants and animals with their physical environment. However, in its present-day context, the term has been widely used as plant ecology, animal ecology and human ecology. Thus, in common usage, there ar three distinct fields of ecological study. However, a new term general ecology, has also been introduced which denotes the fullest extension of the term, and is widely applicable to the vast and complex pattern of the whole universe. Tylore (1936)[5] rightly remarked that "general ecology" is holistic in its treatment and it is the study of "all relations of all organisms to all their environment". As a matter of fact Darwin's concept of "web of life" which provided the logical step for "struggle for existence" was the starting point of ecological approach. Most of the ecologists in the 19th century drew inspiration for their theoretical frameworks, concepts and principles of competition, equilibrium, succession and division of labour from Darwin's concepts. However, the theoretical origins of the ecological approach in social sciences may be traced back to the starting point of Park's[6] ecological approach. Park's paper, "City", was an early stimulus to urban ecology which included Darwin's concept of "web of life". Park's conception of city was based on the principle of "social organism" in which he propounded intimate interrelationships between organism and organisam and between organism and environment. This reference of framework

Park derived from an analogy with the biological world. Park believed that the patterns and relationships evident in a city could be paralleled by land use and people in the cities. The first fundamental principle derived by Park was the concept of competition.

It is human nature to compete for limited space and for access to the most desirable location for his residence and for his business activities. The basic concept was to segregate people and their business activities into relatively homogeneous residential and business suburbs. These homogeneous entities were called "natural areas". Such competitive activities of men in the city were reflected in land values and rental values which segregated the city into the C B D: the areas of commerce and residential areas. Further competition has led to segregation of similar types of people and similar types of business and commercial activities.

The second concept of the traditional ecological theory was that of dominance which was considered one of the fundamental concepts of the plant ecologists. It has been observed that in different plant communities, one species exerts a dominant influence which controls the environmental conditions and in turn encourages and discourages other types of species. Similarly, in cities the C B D plays an important role within the whole complex of the urban areas which exercises control (dominance) over the functional use of land in other parts of the city. The concept of dominance led to the concept of invasion and succession sequences which are documented in the works of Burgess (1925),[7] Cressey (1938),[8] and McKenzie (1925).[9] The traditional ecological studies in a broader sense can be classified into three basic types. The first category includes those studies that deal with the concepts and principles derived from plant and animal ecology to the analysis of the human community, which is largely based on the concepts of competition, dominance and succession. The pioneers of this group are Robert Park[10], Ernest Burgess[11] and R. D. McKenzie.[12] The second group of studies primarily analysed the physical features of specific natural areas which are characterized as social, economic and demographic elements. This group includes the works of Harvey Zorbaugh[13] and Louis Wirth.[14] The third type of studies largely confined themselves to the coding of observable social phenomena, such as crime rates, mental disorders. This group includes the works of Shaw[15], Faris and Dunham.[16]

CRITICISM

The foregoing description clearly shows that ecological research did not take one single line of action. It has been recognized that numerous theoretical and substantive convergences have taken place among the social scientists on the fundamental issues in urban ecology. Now the question arises what were the causes of the downfall of traditional human ecology? The possible answer to this query could be obtained through a series of theoretical and empirical criticisms provided by Alihan (1938),[17] Davie (1937),[18] Gettys (1940),[19] Firey (1945)[20] and Hatt (1946)[21] who have distinctly revealed numerous weaknesses in the traditional ecological theory and have questioned the overall validity of the traditional ecological approach.

THE EMPIRICAL CRITICISM

Human ecologists have made an enthusiastic and productive endeavour in the realm of ecology during the 1920s and early 1930s. However, the late 1930s and 1940s witnessed an increasing criticism of the ecological principles. There was mounting criticism and majority of the scholars directed their attention to uncover spatial correlations. Several scholars discovered a number of weaknesses in the traditional ecological approach. They charged that when they applied the Burgess model of urban growth to Chicago, it was found that various concepts of the ecologists were purely theories. Different sets of natural areas produced overlapping and blurred boundaries which were based on different sets of the ethnic and social data. Firey (1945)[22] provided a most concised fundamental and most widely recognized empirical criticism and identified that sentimental and symbolic elements of place could construct the ecological systems which were governed by competition and the rational allocation of land uses.

THEORETICAL CRITICISM

The theoretical connotations of traditional ecological concepts were strongly criticized by Alihan (1938)[23] who provided the most effective and persuasive theoretical criticism of the ecological position. She pointed out inconsistencies in the use of both natural area and gradient, which, according to her, were mutually exclusive concepts. W. E. Gettys [24] hammered the traditional ecological concepts for the deterministic slant of their approach to human behaviour. According to him there is a sense of bias in their description which was initially drawn by their strong dependence on analogies drawn from the organic ecologists. He suggested that human ecologists should outrightly abandon this dependence on the natural sciences and advised them to focus their attention upon the description, measure analysis and explanation of the spatial and temporal distributions of social and cultural data.

REACTION TO CRITICISM

The mounting criticism of traditional human ecologists had led them to abandon their broad theoretical and empirical studies of community structure. The unfortunate consequence of this criticism was that they shifted their focus and reoriented the analysis of social and cultural determinants of man's location in space and there was a general loss of intellectual confidence amongst ecological writings, even though they continued unabated. Theodorson (1961)[25] prompted the main theoretical readjustment when he incorporated some aspects of cultural facts of urban society into the ecological approach. He was the first scholar to make an attempt to delineate the field of human ecology in a systematic fashion. After the wave of criticism in the late 1930s and early 1940s, various theoretical statements emerged only in the 1950s. However, the works of Quinn (1950)[26] and Hawley[27] were outstanding. Quinn made a distinction between cultural and sub-cultural levels of society. Nevertheless, it was Hawley's statement which made an attempt to delineate the scope of human ecology which has been more suggestive and treated as the starting point for most of the ecological approaches adopted in recent years. He placed more emphasis on the concept of functional relation in the adjustment of human aggregates, their organic and inorganic surroundings. The detailed description of Hawley's analysis postulates that the basis of an ecological organization was differentiation. It was assumed that with differentiation there could not be organization. Hawley proposed that this differentiation was based not only on physio-psychological traits, age, sex and race factors but also determined territorial differentiation.

The contemporary ecologists have used Hawley as their springboard. Duncan[28] in his treatment of the subject matter accepted that Hawley's theoretical exposition represents a consistent, well-elaborated version of the discipline. Indeed, after Hawley's work Duncan and Schnore (1959)[29] were two distinguished scholars who provided a rigorous framework of ecological organization. They added empirical dimensions to the ecological approach which greatly helped scientific analysis of urban structures and functioning of urban communities. Paralleling the sociological experience the 1950s witnessed a theoretical revolution in geography which fostered change in urban geography (Berry, 1965 b.).[30] The advent of the location theory and technical revolution had sparked off quantification in urban geography, as a consequence of which there was a growing overlap between human ecology and modern urban geography.

CONTEMPORARY ECOLOGICAL APPROACH

In the previous discussion it has been well established that traditional human ecology was primarily based upon competition, whereas contemporary ecological analysis is fundamentally concerned with interdependence. The central axis of the contemporary ecological approach is based on the principle of collective adaptation of a population to its environment which is based on a functionally integrated community organization through aggregative and repetitive actions of a large number of individuals. Through basic postulates of the ecological approach it is established that when a community develops an effective organization, it improves its chances of survival in its environment. The contemporary ecological approach has developed along three axioms: (i) Symbiotic and Commensaltic, (ii) Macro-social, and (iii) Interaction of the component parts of social systems. From the ecological point of view, symbiotic and commensaltic interdependences form the basis of social system structure. Symbiotic interdependence is an axiom which postulates structural differentiation and integration of specific roles and functions within a system. The division of labour characterizes the role of symbiotic independence. Commensaltic interdependence primarily concerns itself with supplementary similarities. It attempts to obtain a holistic idea through the co-action of like units which are functionally assembled.

The second axiom of contemporary ecological inquiry is macro-social in perspective. The fundamental principle of this axiom is that social systems exist as entities which demonstrate structural properties that are largely determined by the patterns of organized activities which are the consequence of the routine interaction of various functions performed within the system.

The third axiom of the contemporary approach is based on the assumption which states that the continuous interaction of different component parts of social systems tend to move towards a state of equilibrium Hawley (1968).[31] The state of equilibrium is being achieved functionally, demographically and distributively. The elements of a system try to adjust with prevailing environmental conditions. Thus a stage of stability occurs. This system is referred to by contemporary ecologists as the ecological complex which is constituted of four reference variables as propounded by Duncan (1959).[32] These are: population, organization, environment and technology. However, contemporary ecologists have tried to conceptualize these POET variables. Population pertains to aggregates of human beings who function as a coherent entity. In analysing an ecological complex, ecologists refer to population not only by its individual traits but also include mobility, size, replaceability and interchangeability which show an attempt of a population to adapt to changing environmental conditions. The second important variable of the ecological complex is organization which is synonymous with system. It includes a whole network of symbiotic and commensaltic relationships. It is an attribute of coherent entity. The third key variable, environment, is a generic term, it is a source of sustenance. The fourth denominator is technology which refers to a set of artifacts, tools and techniques employed by a population to adjust to its environment and provide facilities to the organization for its survival. It is a truism that these four key variables are functionally interdependent; any change in the character of one of the four will have serious repercussions on the other three. Thus a majority of contemporary ecologists have tried to focus their attention on explaining interrelationships among demographic, environmental, technological and organizational variables within a social system, (see, for example, Bailey and Mulahy 1972),[33] Kasarda (1971),[34] (1972a),[35] (1972b),[36] (1974)[37], Sly (1972),[38] Schwirian (1974),[39] Frisbic and Poston, Jr. (1975)[40] 1977.[41]

RELATIONSHIP BETWEEN GEOGRAPHY AND ECOLOGY

On an examination of the development of ecological and geographical studies, it is revealed that their approaches to urban problems are similar in many ways. Many of the key concepts of ecology and human geography overlap, for example, the key problem of a contemporary ecologist is to analyse how a population organizes itself and adapts itself to a changing environment. Similarly, modern urban geographers also explain how perceptions of the urban environment determine locational and spatial processes that, in turn, sustain the urban environment. Further, the idea of invasion and succession is reflected in geography by the concept of sequent occupance. The concept of natural region in geography is mirrored in natural areas of ecology. Similarly, the concept of community in ecology has its parallel in the concept of fundamental region in geography. Thus it is evident that ecologists and geographers share basic concepts as a consequence of which many topics studied by geographers and ecologists overlap. Based upon the two shared concepts, functional region and communities, it is but natural that the study of a metropolitan region should be found in both subjects. To further substantiate the overlapping of concepts, it is argued that in geography the voluminous literature on the sphere of influence is in practice not different from the ecological concept of functional independence in an area. Thus, the treatment of geographers and ecologists is identical in virtually all spheres. So Barrows (1923)[42] correctly pleaded that geography should be considered as Human Ecology. Bews (1935)[43] further identifies that a great deal of geographic work comes under the unbrella of human ecology.

ABOUT THE VOLUME

The cultural revolution in geography has greatly affected the nature and subject matter of urban geography and widened its horizons. This volume is designed to show the relevance of ecology to the geographical approach while explaining urban areas. Most of the papers in the present volume confirm that the ecologist's interest in spatial associations and areal patterns is very close to the geographical approach. In essence, in the present volume those articles are included which provide geographical methods of analysing the social structure of towns. The basic idea is to propose an interdisciplinary approach to urban ecological issues. The volume is prepared with a conviction of obtaining a comprehensive knowledge of the patterns, processes and problems of contemporary urban organization and change. The volume introduces the researchers to fundamental assumptions, principles, propositions and paradigms in contemporary ecological inquiry. However, it is not intended to provide an overview of all possible issues and problems of concern in modern ecological inquiry.

The widespread use of the factorial ecology approach denotes that geographers and others possess an appropriate multivariate analytical procedure to monitor the urban residential system. Recently, concern with improvements in the specification of the factor solution on the one hand have strengthened the methodology but on the other have called into question the appropriateness of the method as a vehicle for investigating residential differentiation.

Dennis Conway, in his article, "A Spatio-Temporal Analysis of the Residential System in Port of Spain, Trinidad, 1931-1970", emphasizes that the extension of the conventional factorial approach, namely, spatio-temporal factorial ecology in which residential structure is studied at more than one date, is most appropriate methodology which focusses on how the system has changed. The main concern of the author is the demonstration of the methodological approach to the measurement of residential area changes of an urban system in a Third World Transitional Society, namely, Port of Spain, Trinidad. A spatio-temporal factorial ecological study is conducted to analyse the long- and short-term spatial and temporal alternations in this city's residential structure. Although "colonial" ethnic differentials emerge as dominant discriminators in the residential fabric throughout the period in question, 1931-1970, in the most recent decade 1960-1970 there appears to be the beginning of a reordering of the structure with socio-economic status differentials becoming more important. This evidence supports the tenets of the socio-economic theory which advocates convergence of urban societal order as Third World societies evolve and are incorporated into an expanding world commune.

An examination of the relevant literature shows that a large majority of British factorial ecologists have used either 1961 or 1966 census data as their information source. Few researchers have made an attempt to analyse the potential of the most recent enumeration conducted in 1971. Largely, this reflects the decline in popularity which factor analysis has encountered in the geographical world. Christina R. Victor in her article, "The Urban Dimensions of Greater Nottingham, England", describes the factorial ecology of the Greater Nottingham urban area using 1971 Census information. The author demonstrates that in the present paper technique of factor analysis is used to identify the nature rather than spatial pattern or urban social segregation in the greater Nottingham urban area. In this case study, the study area constitutes a functionally defined, as opposed to administratively designated, urban unit. A comparison of the results with previous British work reveals that it is possible to identify factor structures which are general to the British Urban context. The classic postulates of the school of social area analysis are shown to illustrate little correspondence with these generalizations and the nature of urban social segregation in greater Nottingham in 1971 which could be described in terms of eight primary factors. By employing an oblique rotation procedure, it is demonstrated that these dimensions are intercorrelated. A higher order factor analysis illustrated that these dimensions could be summarized by Three Second Order Dimensions.

A recent review of substantive findings in social area analysis and factorial ecology cites 78 studies involving a total of 66 cities. Of the 78 studies 41 are concerned exclusively with cities in the USA. Only nine studies report factorial ecology of cities in non-industrialized countries. Erling Berge, in his article, "Comparing the Factorial Ecology of Helsinki and Oslo", reports the findings for Oslo and compares the factorial ecology of Oslo with that of Helsinki. According to Berge, the factorial ecology of metropolitan Oslo in 1970 revealed clearly three dimensions. Two were found to be inequality dimensions and labelled Socio-economic Status and Deprivation. The third was found to be equality dimension and labelled Familism. A comparison of the factors in Oslo with the factors found in metropolitan Helsinki in 1960 showed the socio-economic status factors to be the same. But the interpretation of the deprivation factor turned out to be a problem. Comparing the deprivation factor of Oslo with the Familism, Urban residential areas suggested that the factorial interpretation of the Familism Urban factor has overlooked its relation to distribution of material of living. It is suggested that a better label for the factor would be Affluence/Deprivation. The reason for the absence of this factor in the Helsinki study is suggested to be a relative domination of rural/urban differentiation in the data used.

Spatial studies of the internal social formation of British cities have been largely conducted around the theoretical suggestions of social area analysis and the empirical findings of previous factor ecologies. Several researchers have made an effort to identify the social patterns in cities and their relationship to the Bergess and Hoyt models. However, few have investigated these models as generalizations of social change. While these models are broad generalizations, they provide little specific information concerning rates of change and their intra-urban variations. Individual studies of aspects of change in specific cities have been reported for both North America and Britain. However, as yet, little work has been published which has investigated the nature and pattern of social changes in urban social environment at a city circle scale of analysis. Christina R. Victor, in the present paper, "A Multivariate Analysis of Social Area Change in Nottingham C.B. : 1961-71", has made an attempt to use the technique of factor analysis as a means by which to identify change in the structure and pattern of social segregation in Nottingham C.B. The author has investigated two methods of identifying the major procedures and patterns of social area change for 140 enumeration districts in Nottingham C.B. between 1961 and 1971. The dimensions of socio-spatial differentiation are identified using factor analysis in both longitudinal and cross-sectional approaches. The degree of change in the nature of social segregation during this decade is then identified. These dimensions of social change are subsequently compared with the classical social area models and their similarity is assessed.

Since the Chicago sociological school proposed the first interpretation of the urban internal differentiation, studies on social areas within the city have multiplied, accumulating empirical evidences as well as theoretical propositions. Factorial ecology has allowed to test some hypotheses to seek relevance of some factors in the making of urban social areas; some typologies are consistently shown by factor loadings and scores, even if the spatial patterns are widely different. A less inquired case is the social areas patterns with multiple nuclei. This is found in new conurbations formed by a concomitant growth of two or more near cities, whose neighbourhoods come and touch each other. Gabriele Zanetto, in his paper, "Factors and Forms of Social Areas in a Conurbation", identifies that such a problem is evident in the case of conurbation of Venice formed of two main centres. The author in such a case suggests application of a principal component analysis. The author in this paper has made an attempt to analyse the relevance of factorial ecology which he assumes to be significant to test some hypotheses. When applied to Venice the author finds that there are four distinct ecological regions: (i) a central one, (ii) a blue-collar neighbourhood, (iii) historical city, and (iv) the periphery.

Neighbouring, as a concept, has been more ideological assertion than a useful social geographic concept. However, it is evident that there is a level of neighbouring interactions in limited territories. The neighbourhood concept is useful because it does not restrict patterning of relations. Gerald Walker, in his paper, "Social Structure of Neighbouring Working People in West End, Toronto', clearly reveals that neighbourhood, by contrast, though redolent with ideological overtones, has gained some operational precision through empirical study. The author concludes that the examination of data gathered in a working class neighbourhood in Toronto's West End indicates that the respondents live in a number of small social circles, which do not aggregate into a defended neighbourhood. Neighbouring patterns were characteristically based on ethno-religious, near-neighbour groups. Social networks in this undefended neighbourhood served private needs with only diffuse collective functions. Formal associations in the local area appeared to be substitute for articulated personal visiting relations in this safe neighbourhood.

The growth of urban areas in the Western world has been subject to various economic, social and political forces which have resulted in very distinct patterns of segregation or areal differentiation. These forces involve spatial processes and result in spatial patterns which urban students, particularly sociologists, geographers and economists, have incorporated into urban models and theories. Pierre de Vise, in his article, "Forces and Patterns of Segregation and Areal Differentiation in American Cities", provides a brief historical review of human ecology with general explanations of ecological forces and analyses the spatial structure with a view to describing and measuring the ecological forces. De Vise deals with the forces which have resulted in very distinct patterns of segregation or a real differentiation. The abovesaid explanations begin with general ecological forces, proceed to spatial structures, then to description and measurement and end with attempts at new explanations of these forces.

Studies of ethnic residential patterns have been reported for a variety of urban contexts. However, the majority of work has been concerned with the cities of North America. Christina R. Victor, in her paper, "Ethnic Minorities in Nottingham and Areal Differentiation in American Cities", feels that within a specifically British situation fewer studies are in existence. This is especially true for investigations which have sought to review the residential segregation of immigrant groups at more than a single point in time. Of the major British studies, London has been the focus of several investigations and there is a marked lack of published research focussing upon medium sized urban areas and their immigrant communities. Christina R. Victor while preparing her third paper, "Ethnic Minorities in Nottingham C.B." concludes that in this particular investigation the nature and magnitude of ethnic minority group segregation is described. In this paper the author has demonstrated that in Nottingham C.B. considerable variations in residential similarity were identifiable between the major immigrant groups in both 1961 and 1971. The author finds that these variations remained stable and thus suggests that internal cultural differences have influenced the residential distributions of particular ethnic groups. However, the importance of external segregating influences was also demonstrated. The author reveals that in the increasing length of residence of the minority group, these factors decline in importance. Thus whilst it is possible to find evidence to support aspects of both in the internal and external theories of segregation the precise balance between these two sets of factors obviously varies between ethnic groups and their length of residence in the host community.

In the English-speaking countries, factorial ecology and social area analysis more or less lay claim to the field of urban research, especially the spatial organization of urban society. Elizabeth Lichtenburger, in her paper, "The Socio-ecological Structure and Spatial Differentiation of Vienna in the 16th Century", does not follow the conventional socio-ecological structure as propounded in the English-speaking countries. In this paper the author takes a historical viewpoint. Her approach differs basically both in theory and method. She attempts to analyse the spatial and social structure of the mediaeval burghers' town. The author summarizes the functional differentiation of the houses and provides a complete social differentiation of Vienna in the mid-16th century. According to the functional differentiation, the following social strata were recognized: (i) Burghers' houses; (ii) Noblemen's houses; (iii) Houses of clergy and churches; (iv) Public buildings town; (v) University dormitories.

Despite numerous studies on racial residential segregation, the relative importance of various factors determining variations in the levels of segregation among cities has not been fully analysed. Relatively more research efforts have been directed to explaining intra-urban patterns of segregation and most of the existing studies rely heavily on 1960 and pre-1960 data. The persistent trends of residential segregation in American cities over the 1960-70 period points to the need for an analysis of causes of segregation. Gill C. Lim, in his paper, "Residential Segregation: An Empirical Analysis of Intercity Variation", examines the determinants of residential segregation by race using cross-sectional data for 43 cities in 1970. It differs from previous studies of segregation on two major aspects. First, the paper is concerned with an analysis of intercity variation in residential segregation. A series of empirical models are developed and estimated to assess the relative importance of economic and attitudinal factors affecting variations in the level of segregation among metropolitan areas. Second, It provides an analysis of both black-white and non-white segregation. The major efforts of the author in the present paper are that he has evaluated two separate sets of models to observe major similarities and differences between black-white and non-white-white segregation. The author generates a series of segregation models in order to evaluate the relative importance of economic, attitudinal and demographic determinants of segregation. The empirical analysis presented in this paper suggests that economic factors contribute little to the explanation of intercity variations in the level of racial segregation. It also identifies that the regional differences in attitudes and other factors not explained by economic and demographic variables appear to have a significant impact on residential segregation. Third, while it is difficult to establish a clear causal relationship between segregation and living conditions, the result implies that residential integration has possible association with high prices in the short run. It appears from the analysis that demographic factors such as population size, relative size of black population and relative growth rate of non-whites do not seem to have substantial effects on segregation. The author also postulates that differences in effect of the regional variables between black-white and non-white-white segregation are observed.

Throughout the world, increasing urban populations give rise to undifferentiated settlement patterns with a pot pouri of buildings and groups of people without an understanding of the underlying principles. As a result, effective improvements in urban environments and the provision of community needs cannot be achieved. Against this background E. Van Loggerenberg and J. N. Scheepers identify and describe objectively the creation of specific residential patterns. In their paper, "Unique Residential Areas in Metropolitan Witwatersrand: Towards Interpreting Multi-Dimensional Outcomes", they point towards: (i) the identification of dimensions from a multivariation of data which are fundamental to the important characteristics of the population. This means that the actual statistical variations should be technique-independent to explain the real-world situations: (ii) the identifications of spatial patterns that result from the abnormal problems of unique factors or dimensions. The authors illustrate that misinterpretation of analytical results may lead to conclusions that cannot withstand the test of reality.

The works of the Chicago School sociologists and some comparatively recent ones by geographers have shown that the social dimension of geographic space is not the product of chance but rather is the result of a certain number of fairly clearly defined parameters, such as socio-economic status and family status. It is a truism that people's residential address reflects, in a general sense, the basis of a social stratification of urban space and this social segregation is defined in relation to a number of different factors, such as education and economic levels, ethnic origins and family size. Jean-Pierre Thouez and Peter M. Foggin, in their article, "The Social Ecology of Cardiovascular Disease and Mortality" demonstrate that socio-economic categories, however, cannot be viewed as identical to the concept of social class. Nonetheless, whatever measurement tool is used, the spatial analysis of differential mortality in its relationship to spatially segregated social stratification almost invariably reveals an inverse relationship between risk of death and social hierarchy. The authors, in this paper, place emphasis on the social ecology of disease and mortality in the urban environment, leaving to one side physical factors, even though the two realms are clearly related in time and space. The following presentation is threefold: first, a brief discussion of research bearing on social differences in relation to death statistics; second, an analysis of mortality and morbidity data for Montreal Islands; and, third, some methodological discussions in relation to the geographic study of illness and death.

While settlement planning in socialist societies has pursued a number of common goals, the cultures and the environments within which such policies have been attempted have been remarkably diverse. The combination of centralized economic planning and public control of the means of production has permitted large-scale transformations to take place in such societies, although individual national differences in pre-existing settlement forms, demographic structures and levels of economic development have varied the impact of those transformations from one country to the next. Derek R. Hall, in his article, "Emerging Residential Structures of Socialist Cities", has made an attempt to reveal the emerging residential structure of socialist urban spatial set-up and optimum city size. He provides a vivid picture of stringent land use zones and rationalized internal movements. The author also explores the nature and role of city centres, planned development of residential areas, and housing tenure. The author also identifies emerging ecological patterns and propoundes a model of factors affecting social segregation in socialist cities.

In any country the changing spread of population density over time is an indicator of its growth and urbanization. Ashok K. Dutt and Gerardine D'Sa, in their article, "Density Gradients of Calcutta City 1872-1981", efficiently assess the spatial patterns of population density in Calcutta, and analyse these patterns in the context of its historical roots, socio-economic status and technical association during the past 110 years. The purpose of the authors in this paper has been to study Calcutta's density gradients historically and find out whether Newling's youthful stage can be identified from the earlier spatial patterns. They conclude that basically the pre-independence pattern has been retained after independence as the centre continues to be dominated by commercial and government-related activities. Finally, they conclude that a city with a colonial base like Calcutta possibly always had a low central density compared to its immediate ring.

It has long been evident that the context in which people live affects how they live. Living in high status communities rather than in low status communities influences voting behaviour, social and political attitudes, economic status and the nature of social networks. These effects are best understood in the case of voting, where strong circumstantial evidence suggests that when one social class is dominant in a constituency a larger proportion of the electorate votes in accordance with that class's predisposition but contextual effects do not stem only from socio-economic status. The ethnic composition of a community affects the attitudes to racial and ethnic questions, individual modernity, earnings of both majority and minority groups and income in equality. Other influences, for example, are on political behaviour, income, migration, modernization and propensity to strike. Research on contextual effects in Australia has been severely limited mainly due to lack of appropriate ecological data and Australian social scientists have ignored their potential. The paper by Ian McAllister and Jonathan Kelley, "Contextual Characteristics of Australian Federal Electorates", seeks to remedy that neglect. The authors in their article postulate that social science has propounded two types of variables to account for social and political behaviour: on the one hand these are individual characteristics, such as social class, age, sex and ethnic origin; on the other, there are contextual characteristics, for example, the class milieu, the urban, suburban and ethnic composition of the areas in which people live. Observed patterns of behaviour result from the interaction between these individuals and contextual characteristics. Drawing on data from the 1976 Census, the paper examines the contextual characteristics of Australian federal electorates. The paper aims, firstly, to isolate the main structural dimensions of federal electorates, using factor analysis to distil the wider range of census variables into a small number of conceptually and statistically unambiguous scales, each consisting of a number of individual census variables. Four dimensions are delineated: ethnicity, socio-economic status, family and urban-rural differences. Secondly, the paper aims to provide data for each of these scales, enabling other researchers to easily and routinely incorporate this information into existing data sets and to analyse it using straightforward techniques.

REFERENCES

1. Barry Commoner, **The Changing Circle of Nature, Man and Technology**, (1971) p. 39.
2. Neol P. Gist and Sylvia Fluis Fava, 1974, **Urban Society**, Thomas Y. Crowell Company, New York.
3. Hawley, **Human Ecology: A Theory of Community Structure.**
4. Brian J. Berry and J.D. Kasarda, **Contemporary Urban Ecology**, Macmillan Publishing Co. Inc., New York (1977).
5. W.P. Tylore, "What is Ecology and What Good is it Ecology", XVIII, 1936, 335.
6. Park, **Human Communities**, Free Press, New York, (1952) p.14.
7. Burgess, "The Growth of the City: An Introduction to a Research Project" in Robert Park, Earnest Burgess and R.D. McKenzie (eds.), **The City**, University of Chicago Press, (1925) pp. 47-62.
8. Cressey, "Population Succession in Chicago, 1898-1930", **The American Journal of Sociology**, (1938) 44, 59-61.
9. McKenzie Roderic, "Ecological Succession in the Puget Sound Region", **Publications of the American Sociological**

Society, (1929) 23, 60-80.
10. R. Park, op. cit.
11. Burgess, op. cit.
12. McKenzie, op. cit.
13. Harvey Zorbaugh, "The Natural Areas of the City", **Publications of the American Sociological Society,** (1926), 20, 188-197. Reprinted in George A. Theodorson (ed.), **Studies in Human Ecology,** pp. 45-49, NewYork. Harper and Raw, (1961).
14. Louis Wirth, "Urbanism as a way of Life", **American Journal of Sociology,** (1938), 44, 18-19.
15. Shaw Clifford, F.M. Zorbaugh, Henery McKay and Leonard Cottrell, **Delivevey Areas,** University of Chicago Press, Chicago (1925).
16. Faris, Robert E.I. and H. Warran Dunham, **Mental Disorders in Urban Areas,** University of Chicago Press, Chicago (1939).
17. Alihan, Milla, **Social Ecology: A Critical Analysis,** Columbia University Press, New York (1938).
18. Davie, Maurice R., "The Patterns of Urban Growth" in George P. Mardock (ed.), **Studies in the Science of Society,** Yale University Press, New Haven, (1937) pp. 133-161.
19. Gettys, Warner E., "Human Ecology Theory" **Social Forces,** (1940), 18, 469-476.
20. Firey Walter "Sentiments and Symbolism as Ecological Variables", **American Sociological Review,** (1945) 10(2), 140-148.
21. Hatt Paul, "The Concept of Natural Area", **The American Sociological Review,** (1946) 11(4), 423-427.
22. Firey Walter, (1945) op. cit.
23. Alihan, Milla, op. cit.
24. Gettys, op. cit.
25. Theodorson, George A., (ed.), **Studies in Human Ecology,** Harper and Row, New York (1961).
26. J.A. Quinn, **Human Ecology,** NewYork (1950).
27. Hawley, A.H., op. cit.
28. Duncan, O.D., **Human Ecology and Population Studies.**
29. Duncan and Schnore, "Cultural Behavioral and Ecological Perspective in the Study of Social Organization", **American Journal of Sociology** (1959), LXV, 132-146.
30. Berry, Brian, J.L., "Internal Structure of the City", **Law and Contemporary Problems,** (1965a) 30 (1), 111-119.
31. Hawley, "Human Ecology" in **International Encyclopaedia of Social Sciences** (1968) Vol. 4, pp. 328-337, McMillan, NewYork.
32. Duncan, Ottis Dudley and Lad Schnore, "Cultural, Behavioral and Ecological Perspectives in the Study of Social Organization", **American Journal of Sociology** (1959) 65, 132-146.

33. Bailey, Kenneth D. and Patrick Mulahy, "Socio-cultural Versus Neo-classical Ecology: A Contribution to the Problem of Scope in Sociology", **The Sociological Quarterly** (1972), 13, 37-48.
34. Kasarda, John D., "Economic Structure and Fertility: A Comparative Analysis", **Demography** (1971), 8, 307-317.
35. Kasarda, John D., "The Impact of Suburb and Population Growth on Central City Service Functions', **American Journal of Sociology** (1972a), 77 (6), 1111-1124.
36. Kasarda, John D., "The Theory of Ecological Expansion: An Empirical Test", **Social Forces** (1972b) 52, 165-175.
37. Kasarda, John D., "The Structural Implications of Social System Size: A Three Level Analysis ", **American Sociological Review**, Feb. (1974) 39, 19-28.
38. Sly, David F., "Migration and Ecological Complex", **American Sociological Review** (1972), 37, 615-628.
39. Schwirian, Kent P., "Some Recent Trends and Methodological Problems in Urban Ecological Research" in Kent P. Schwirian (ed.), **Comparative Urban Structure: Studies in the Ecology of Cities** (1974) pp. 3-31. Lenington, Mass., D.C. Heath and Company.
40. Frisbic, W. Parker and Dudley L. Poston, Jr. "Components of Sustenance Organization and Non-metropolitan Population Change: A Human Ecological Investigation", **American Sociological Review**, Dec. (1975), 40, 773-784.
41. Frisbic, W. Parker and Dudley L. Poston, Jr. "Sustenance Organization and Population Redistribution" in **Non-metropolitan America**, Sage Publications, Berkely Hills, Calif (1977).
42. Barrows, "Human Ecology", **Annals of the Association of American Geographers** (1923), XIII, 1-47.
43. J.S. Bews, **Human Ecology**, (1935) London.

TWO

Dennis Conway

A Spatio-Temporal Analysis of the Residential System in Port of Spain, Trinidad, 1931-70

INTRODUCTION

Widespread use of the factorial ecology approach may suggest that geographers and others possess an appropriate multivariate analytical procedure to monitor the urban residential system. Recently, concern with improvements in the specification of the factor solution on the one hand have strengthened the methodology (Chojnicki and Czyz, 1976; Conway and Haynes, 1977; Davies, 1978; Giggs and Mather, 1975; Johnston, 1973b), but on the other hand have called into question the appropriateness of the method as a vehicle for investigating residential differentiation (Johnston, 1979; Newton and Johnston, 1976; Perle, 1977, 1978). While the use of alternative methodologies such as multidimensional scaling and Semples entropy statistic in the identification and classification of districts is a healthy sign (Johnston, 1979; Klaff, 1973; Semple, Youngmann and Zeller, 1972), still it can be argued that factorial ecology is an appropriate technique when the goal is systematic data exploration and description and the concern is how the urban residential system coheres (Berry, 1971; Clark, et al., 1974; Davies, 1978). The particular concern here is with the extension of the conventional factorial ecology approach, namely spatio-temporal factorial ecology in which the residential structure is studied at more than one date and the analysis focusses on how the system has changed (Johnston, 1976).

Perhaps the most formidable obstacles to this obvious improvement in the factorial ecology approach are operational considerations of data compatibility. Census tract data (the ubiquitious source for most ecological analysis) is notoriously incompatible between censuses. Variable definitions change, census tract boundaries are rarely consistent over time, and an everpresent problem is the fact that as the urban system expands, new residential subareas emerge in a later census enumeration which have no predecessors. These problems alone may be sufficient to explain why so few spatio-temporal factorial ecologies have been attempted, but the roll-call of methodological difficulties associated with the measurement of residential area change is far from exhausted. Even if data compatibility is achieved between dates in longitudinal studies, the structuring of the chosen indices of change may have a subjective influence on the ensuing analysis (Johnston, 1973a; Murdie, 1969). Questions of solution parsimony, factor invariance and variable homogeneity must be satisfactorily addressed (Conway and Haynes, 1977; Davies, 1978; Giggs and Mather, 1975; Haynes, 1971). Furthermore, it would appear essential to include in an analysis of change within a finite time period an account of the long-term changes that are in progress and are affecting the residential system at the initial date as well as the shorter-term changes occurring during the time period under scrutiny (Brown and Horton, 1970).

The last point is of particular relevance for the urban system chosen for this study. Almost without exception, spatio-temporal factorial ecologies have been conducted on urban residential systems in "advanced" industrial or post-industrial societies; in the United States (Brown and Horton, 1970, Hunter, 1971; Sweetzer, 1962), in Canada (Greer-wootten, 1972; Haynes, 1971; Murdie, 1969), and in Australia (Johnston, 1973; Timms, 1970). To date, there has been only one documented analysis of residential area change of an urban system in a Third World transitional society, a comparative statics factorial ecology of Hong Kong (Lo, 1975). Port of Spain, Trinidad, the urban system under investigation here, is another representative of the Third World and it would appear that in recent decades the city may have undergone considerable transformation as the society has modernized from its colonial to neo-colonial state (Cross, 1979; Oechsli and Kirk, 1975; Rampersad, 1962). Therefore, not only is the analysis of changing patterns of the residential system in such a city a valuable contribution to the body of comparative urban ecology research (Masotti and Walton, 1976), but in addition, the methodological approach used here measures long- and short-term spatial and temporal changes.

Some earlier attempts to isolate dimensions of residential area change have suffered from a lack of consistency in their choice of indices (Brown and Horton, 1970; Murdie, 1969; Sweetzer, 1962). Another deficiency has been that reliance was on visual interpretations of spatial patterns of change from comparisons of separate factorial ecologies at two different dates (Hunter, 1971; Lo, 1975; Timms, 1970). Fortunately, available data from Trinidadian censuses are sufficient to allow an analysis of changes in the residential system of Port of Spain between 1960 and 1970 to be integrated with an analysis of the evolving urban ecological structure from 1931 to 1970.

In brief, the combination of analytical procedures of this spatio-temporal factorial ecology are as follows. First, the temporal evolution of Port of Spain's ecological structure between 1931 and 1970 is established. Explicitly aspatial in approach, this part of the analysis tests the dimensions of residential differentiation for increasing independence and stability, in accordance with an existing evolutionary model of residential area differences (Conway, 1976; Timms, 1971; Johnston,1976). Comparative statics analyses for the periods 1931 to 1960 and 1960 to 1970 identify the nature and character of changing dimensions and are complemented by a confirmatory longitudinal test using canonical correlation analysis which specifically identifies persistent elements in the urban ecological structure between 1960 and 1970. Following this investigation of the long- and short-term changes in the societal fabric of the city, the spatial patterns of the changing residential mosaic during the 1960-70 decade are analyzed. From an orthogonally-constrained comparative statics analysis of the ecological structure at the two dates, 1960 and 1970, a classification of changing residential areas is derived. This classification procedure is based on a method used in a Montreal factorial ecology study (Greer-Wootten, 1972). Fundamentally, a classification of stable and unstable residential areas is developed by ranking the factor scores of the two prinicipal orthogonal dimensions into sextiles for the two dates, 1960 and 1970, and concentrating on the extremes (positive and negative) of the scores. The degree of movement of sub-areal units between sextile classes from 1960 to 1970 is used to identify stability and relative change. Utilizing this classification, a map of the spatial pattern of changes in residential area character of Port of Spain between 1960 and 1970 can then be constructed.

Omitted from this brief outline of the study's procedures is reference to a series of reliability tests that provide stricter control in operational criteria when using factor analysis for spatio-temporal factorial ecology and the analysis of change. An earlier paper has addressed this issue (Conway and Haynes, 1977). Furthermore, it is accepted that use of the factorial ecology approach carries with it the limitation inherent in urban ecological study, namely the assumption that the system under scrutiny comprises a set of entities interacting and interdependent among themselves and with their environment (Berry,

1971). However, given that the measurement of change is notoriously difficult (Harris, 1963; Schilderinck, 1970) and given the nature of the urban residential system in all its complexity and dynamism (Johnston, 1971), it would appear appropriate to begin the task of unravelling that dynamic via the use of spatio-temporal factorial ecology.

The emphasis in this study is the demonstration of a methodological approach to the measurement of change in the urban residential system. This is not to deny the importance of a position which calls to "aspire to generality by attempting to be comparative" (Berry, 1971). In fact, to pursue the exploratory analysis without recourse to existing theoretical statements would be a self-limitation that is overly cautious. Accordingly and given recent calls for a continuation of effort in comparative urban ecology research (Abu-Lughod, 1975; Masotti and Walton, 1976; Walton, 1976), the analysis which is conducted here, pursues a course which contributes to that theoretical orientation (Hawley, 1971; London and Flanagan, 1976; Schwirian, 1974).

A Theoretical Position on Changes in Urban Ecological Structure

Theory on the ecological structure of cities recognizes linkages between social, economic and political processes and the spatial manifestations of social structure as location patterns in the residential space of the city are transformed. A recent formulation of an evolutionary theory of urban social spatial structure is by Timms (1971). Building on social ecology explanations of urban residential differentiation (Wilson and Wilson, 1945; Shevky and Bell, 1955; McElrath, 1968), and largely by inductive reasoning, Timms theorizes upon the basic dimensions of societal organization and their separation through time as cities modernize (Timms, 1971; Johnston, 1976). Concomitantly there appears to be empirical support for Timms' embryonic theory in comparative factorial ecology studies (Abu-Lughod, 1969; Berry and Rees, 1969; Briggs and Conway, 1975; Johnston, 1976; Schwirian and Smith, 1974; Timms, 1970). Other commentaries on the evolution of urban ecological structure have drawn on a broad range of comparative studies in countries throughout the Third World and essentially arrived at a similar general conclusion (Hawley, 1971, London and Flanagan, 1976). And, irrespective of the problems attending the interrelationships between urbanization, industrialization and modernization there appears a concensus of support for the generalization that urban systems may be undergoing convergence (Hawley, 1971; Timms, 1971). The following quote presents the core of this general view that nations and their evolving urban systems are being absorbed into an expanding world system (Hawley, 1971, 311-3):

> As one society after another has yielded to the superior economic and political power of an industrialized market economy, it has been thrust along a path toward a drastic internal reorganization... Participation in the world economy has demanded of every society the adoption of a new technological regime, a reorientation of resources, a fundamental reconstitution and realignment of social units and a broad-scale redistribution of its population. Urbanization is more or less localized manifestation of this regionwide or multiregional transformation... Although urbanization begins in very different cultural contexts, in each instance the trend soon begins to reproduce phases and patterns that have occurred in other times and places. The convergence tendency penetrates surface features, such as spatial arrangements... to affect family structure, types of voluntary associations and interpersonal relations of virtually every kind.

Drawing on this admittedly loosely-woven rubric which sketches out a theory of comparative urbanization based on forms of economic integration and particularly, capitalist accumulation (Harvey, 1973; Lojkine, 1972; Timms, 1971; Walton, 1976), three fundamental propositions can be proposed regarding the character of societal change and the ensuing evolution of urban ecological structure.

1. Irrespective of the transitional state of society, the residential structure of all cities is potentially differentiated by four main "latent constructs" (i) socioeconomic status, (ii) life-style (or family status), (iii) ethnicity, and (iv) migration status.

2. The dimensions of socioeconomic status and family status are the basic and dominant constructs and the extent to which they are independent, either of each other or of ethnicity and migrant status associations reflects the stage in economic and technological transformation.

3. The relationship of ethnicity and migration status to the main "latent constructs" of socioeconomic status and family status reflects the cultural context within which modernization transformations occur.[1]

With reference to Port of Spain, Trinidad we are dealing with an urban system which has developed as an integral actor in the transformations of the Trinidadian society from its colonial to neo-colonial state (Ayub and Cruickshank, 1977;

Ramsaran, 1977). During the colonial era, a particularistic-ascriptive social structure evolved in trinidad in which a small white ethnic minority dominated an ethnically distinct majority of negroes, East Indians and Chinese (Braithwaite, 1953; Lowenthal, 1972; Smith, 1965). The change of the flag in the early 1960's formally altered Trinidad's relationship with its metropolitan parent (Ryan, 1972), but it may be anticipated that alterations in its social structure were slower to accomplish (Beckford, 1972; Best, 1968; Clarke, 1973; Fanon, 1967). However, recent commentary on societal change in West Indian societies has suggested a reconstitution of the social and spatial formations during the neo-colonial period as the spread of universal education, the transfer of political power to Trinidadian nationals and the increasing access of Creoles and non-whites to occupations in previously restricted sectors has led to a slow, but persistent, breaking-down of the former racially-dominated stratification (Clarke, 1975; Goodenough, 1975; Lowenthal, 1972). Whether this anticipated "convergence" can be observed in the social and spatial formations of Trinidad's primate city is the general question addressed in this spatio-temporal factorial ecology.

Temporal Changes in Port of Spain's Ecological Structure, 1931-70

An investigation of the character and relationships between the basic dimensions of residential differentiation during the 1931-70 period will provide evidence of the long- and short-term changes in Port of Spain residential system in order to determine the validity of the convergence thesis. The analysis is in three stages.

(a) Image factor analysis with Oblimin (oblique) rotation is used on conformable 1931 and 1960 data sets to establish the aspects of dynamic change in dimension relationships during that thirty-year period.

(b) Image factor analysis with Oblimin (oblique) rotation is used on a larger set of conformable 1960 and 1970 data sets to investigate the continuation in trends of dimension relationships.

(c) Canonical correlation analysis is applied to comparable data matrices for 1960 and 1970 to identify the continued existence in 1970 of relationships between variable clusters already evident in the 1960 ecological structure.

In order to ensure direct comparability between time periods in the 1931-60 test a restricted operational urban area is necessary. The 53 subareal units comprising the 1931 city are adopted as the base urban system and the 1960 enumeration districts are organized to conform to their 1931 equivalents (Fig. 2.1). Because of the need to ensure variable comparability between successive census enumerations, and constrained by the rather restricted set of variables available in the 1931 census, it is possible to use only 16 variables in this first longitudinal analysis (Table 2.1).

Particularly unfortunate is the inability to incorporate measures of occupation or migration status. However, direct comparability between enumeration districts and enumeration schedules in the two recent censuses of the city of Port of Spain, allows a more comprehensive urban area to be used as the base system in the 1960-70 longitudinal analysis (Fig. 2.2). Concern with the structure of the data set so that a balance is achieved between variables representing socioeconomic family status and ethnicity and migrant status (Davies, 1978), is tempered by the existence of comparable measures enumerated in the two censuses. A lack of representative occupational data in the 1970 enumeration schedule meant that the study had to rely on educational status categories disaggregated by sex to serve as substitutes for occupational categories. Furthermore, the concern for comprehensive representation raises without presenting a totally satisfactory solution, the problem that 'closed number sets' brings to factorial ecology study (Johnston,1979). In this study, then, it is felt that a compromise has to be reached, which trades off analytical objectives and methodological queries, and yet generates analysis which is informative. Accordingly, the variable set is made up of 30 comparable variables indexing as wide a range as possible of the attributes of the residential population. In addition to these population characteristics, three spatial variables, distance, area and population density are included.

In the two longitudianl analyses, the residential structure under investigation is only part of the urban system, namely that entity which is within the boundary of the city of Port of Spain, and for which there exists a continuity of census enumerations. Omitted from the longitudinal analyses are areas outside the city boundaries which have become residential areas in the latest decade, but were non-existent in prior enumerations. A cross-sectional analysis of the 1960 ecological structure of the larger metropolitan area of Greater Port of Spain was conducted and compared to the restricted 1960 city structure used in this study (Conway, 1976). Although there were some differences in the amount of variance comparable dimensions explained in the two ecology studies, the use of the

TABLE 2.1

Variable Comparison: 1931-1960 Port of Spain

1931 PORT OF SPAIN	RELATIONSHIP	1960 PORT OF SPAIN
Sex ratio	equal	Sex ratio
Adult married men(%)	opposite	Adult males never married (%)
Population aged 0-2 (%)	equal	Population aged 0-2 (%)
Population aged 3-14 (%)	equal	Population aged 3-14 (%)
Population aged 65+ (%)	equal	Population aged 65+ (%)
Native Trinidadian population (%)	part of	'Mixed' population (%)
East Indian population (%)	equal	East Indian population (%)
Barbadian born population (%)	part of	Negro population (%)
White population (%)	equal	White population (%)
Chinese population (%)	equal	Chinese population (%)
Population of Roman Catholic religon (%)	equal	Population of Roman Catholic religion (%)
Population of Protestant religions (%)	equivalent	Population of Anglican religion(%)
Population of Hindu religion(%)	equal	Population of Hindu religion (%)
Population of Muslim religion (%)	equal	Population of Muslim religion (%)
Adult male illiterate population (%)	opposite	Adult males % with university education
Adults who can read and write(%)	equivalent	Adult females % with school certificate

restricted urban system for this study's comparison through time appears a reasonable operational judgement. Centainly this longitudinal analysis can establish the pattern and variation in the resultant transformations of the residential system of the old existing city.

ANALYSES

Image factor analysis with an Oblimin (oblique) rotation was applied to the 1931 and 1960 data sets to examine the character of, and the relationships between, the basic dimensions of Port of Spain's residential differentiation (Tables 2.2 and 2.3). Although the comparison of factor patterns is difficult where there have been some changes in variable definitions between the censuses, there appears to be considerable similarity in the dimensions which emerge for 1931 (Table 2.2) and 1960 (Table 2.3).[2] Ethnic-religious variation is a dominant feature of residential differentiation in both ecological structures. The major ethnic groups, Negro and White, each have distinct socioeconomic status and family status associations, and the East Indian minority is represented by a distinct dimension. The major difference between the two structures, the presence in 1931 and absence in 1960 of a Bajan (Barbados-born immigrants) dimension, follows from the consolidation of Trini (Trinidadian-born) with Bajan into a Negro category for the 1960 census.[3]

Comparison of the inter-factor correlation matrices in Table 2.4 shows that the family status dimension (Factor 5, column 5), has moved towards greater separation. Three of its four correlations with other dimensions are lower in 1960 than in 1931. It would appear that separation has occured over time. However, a comparison between the 1931 and 1960 relationships of the family status/Trinidadian and family status/Negro factors with other dimensions shows little decline in interdependencies. And, increased rather than reduced interdependence between the White dimension and other ethnic dimensions appears to have occurred by 1960 (Table 2.4). In sum, comparison of the inter-factor correlation matrices in Table 2.4 shows there has been a limited separation of the dimensions of residential differentiation from 1931 to 1960. With dimensions dominated by ethnic and religious associations, there is little evidence of a general reordering of the residential structure in Port of Spain during the thirty year period.

In the 1960-70 longitudinal analysis a more extensive urban area and a larger number of variables are used than in the 1931-60 comparison. However, there is a strong similarity between the 1960 factor pattern based on this more comprehensive

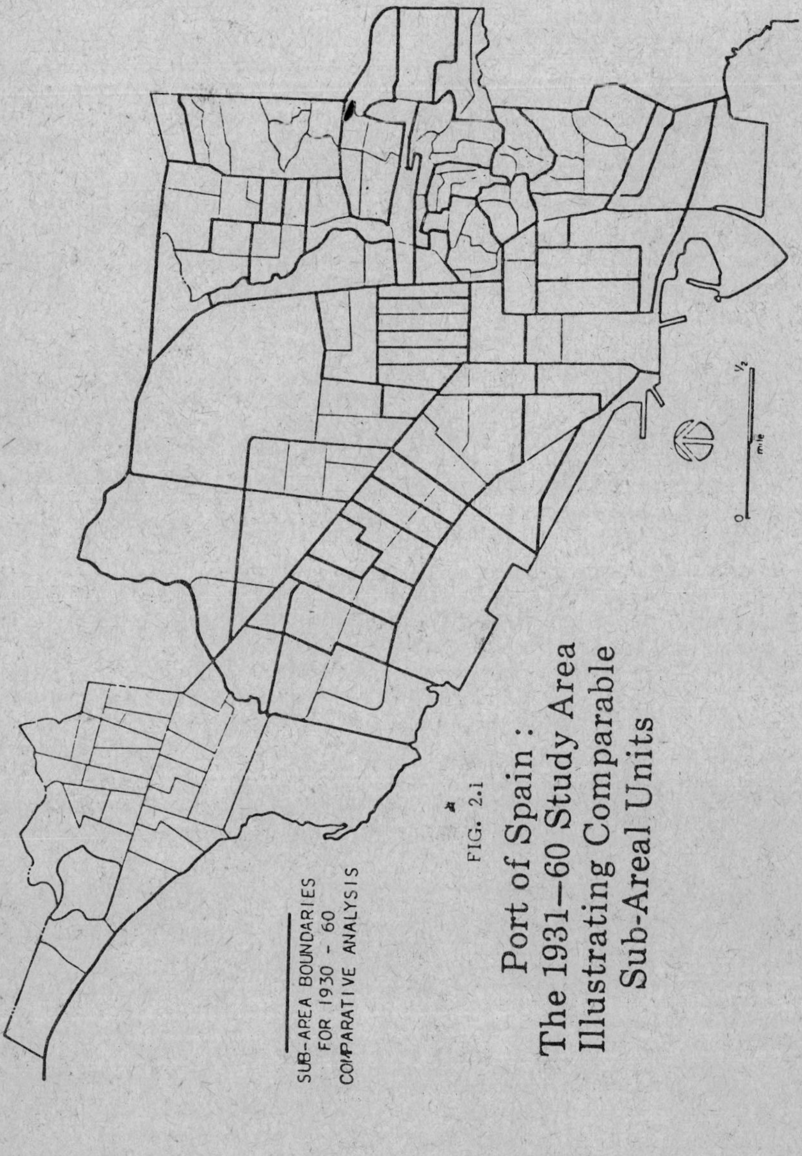

FIG. 2.1

Port of Spain :
The 1931–60 Study Area
Illustrating Comparable
Sub-Areal Units

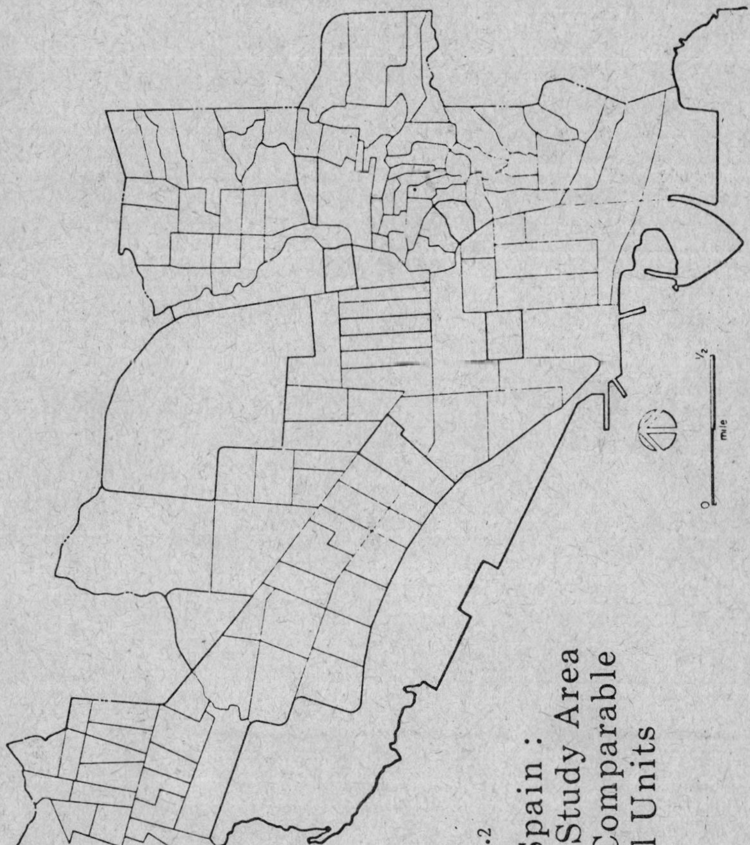

FIG. 2.2

Port of Spain :
The 1960–70 Study Area
Illustrating Comparable
Sub-Areal Units

TABLE 2.2

1931 Port of Spain: Image Analysis With Oblimin Rotation Factor Pattern Loadings

	1 Fam/Trini	2 E. Indian	3 White	4 Bajan	5 Family	6 Chinese*
Sex Ratio	.3592				.3540	-.4122
Adult married men (%)	.4217			.3649		
Population aged 0-2 (%)					.8043	
Population aged 3-14 (%)	.6340					
Population aged 65+ (%)	.8459					
Native Trinidadian Population (%)	.6932					
East Indian population (%)		.9028				
Barbadian born population (%)				.9798		
White population (%)			.8634			
Chinese population (%)						-.8001
Population of Roman Catholic religion (%)	.8306					
Population of Protestant religions (%)				.7233 (.2560)		
Population of Hindu religion (%)		.7093				
Population of Muslim religion (%)		.9801				
Adult male illiterate population (%)	.3202	.5468			.3114	
Adults who can read and write (%)	.6347					

* Factors are labelled on the basis of the high loadings of contributing variables. Loadings 0.3 are excluded.

TABLE 2.3

1960 Port of Spain: Image Analysis with Oblimin Rotation
(Restricted Data Set)

	Factor Pattern Loadings				
	1	2	3	4	5
	Fam/ Negro	White/SES	E. Indian	Chinese	Family*
Sex Ratio					.6134
Adults never married (%)	.9401				
Population aged 0-2 (%)	.4566				
Population aged 3-14 (%)	.6992				
Population aged 65+	.7529				
Negro population (%)	.9436				
Mixed population (%)	.4478			-.4091	-.3044
East Indian population (%)			.9550		
White population (%)		.9588			
Chinese population (%)				-.8170	
Population of Roman Catholic religion (%)	.4233				
Population of Anglican religion (%)	.9900				
Population of Hindu religion (%)			.7636		
Population of Muslim religion (%)			.6616		
Adult males, % with university education		.9718			
Adult females, % with school certificates		.3059			

* Factors are labelled on the basis of the high loadings of contributing variables. Loadings 0.3 are excluded.

TABLE 2.4

Oblique Structure Correlation Matrices*

1931

	Fam/Trini	E. Indian	White	Bajan	Family	Chinese
Fam/Trini	1.000	-.378	.009	.468	.503	-.111
E. Indian		1.000	-.268	.152	.456	-.718
White			1.000	-.104	-.273	.128
Bajan				1.000	.409	.012
Family					1.000	.351
Chinese						1.000

1960

	Fam/Negro	E. Indian	White	[ ]	Family	Chinese
Fam/Negro	1.000	.491	-.143	---	-.352	-.306
E. Indian		1.000	-.028	---	-.056	-.327
White			1.000	---	-.314	-.359
[ ]				---	---	---
Family					1.000	.311
Chinese						1.000

* Factors are identified by their labels (See Tables 2.2 and 2.3)

set of variables (Table 2.5) and the 1960 factor pattern based on the restricted data set (Table 2.3). The main difference, the lack of a mobility dimension in the restricted set (Table 2.3) can be attributed to the absence of representative variables.

Apart from some changes in the order of the dimensions extracted, there does appear to be a general consistency in the emergent latent structures of residential differentiation in the 1960 and 1970 factor patterns (Tables 2.5 and 2.6). The following six dimensions can be identified in the 1960 factor solution:

(1) Family status, Negro group

This is characterized by a very high loading (0.99) on the Negro variable. This ethnic group is associated strongly with

TABLE 2.5

1960 Port of Spain: Image Analysis with Oblimin Rotation

Factor Pattern Matrix[a]

Variables	1 Family /Negro	2 Socio-economic Status/ White	3 East Indian	4 Family	5 Mobility	6 Socio-economic Status/ Ethnic
1. Sex ratio				.9800		
2. Population aged 0-2 (%)	.8853					
3. Population aged 3-14 (%)	.8317					
4. Population aged 65+ (%)	.4188				-.4267	.3953
5. Adults never married (%)	.6630				-.4079	
6. Person/household ratio				-.6182		
7. Live births, % per adult females at risk (aged 15-45)				.3930		-.3520
8. Adult females, % with 4 or more children	.3446					-.4510
9. Household-head sex ratio		-.7167				
10. Adult males, % with secondary education						.8811
11. Adult males, % with school certificate						.7015
12. Adult males, % with university education		-.7683				

Table 2.5 continued....

	1	2	3	4	5	6
13. Adult females, % with secondary education						.8228
14. Adult females, % with school certificate		-.4050				.7206
15. Adult females, % in sales and clerical professions						.9027
16. Adult females, % in professional occupations						.6307
17. Negro population (%)	.9945					
18. White population (%)		-.7440				
19. East Indian population (%)				.9451		
20. Chinese population (%)					-.3766	.5277
21. 'Mixed' population (%)						.7141
22. Population of Anglican religion (%)	.9369					
23. Population of Roman Catholic religion (%)	.6109					
24. Population of Hindu and Muslim religions (%)			.8478			
25. Native Trinidadian population (%)	.7884					
26. Migrant population (%)	.5801				-.3645	
27. Nonmover population (%)	.8077					
28. Density		.3174				
29. Distance from CBD			.3389		.7392	
30. Area		-.8270				

a--Loadings 0.30 are excluded. The principal loadings used to identify the dimension character are underlined.

TABLE 2.6

1970 Port of Spain: Image Analysis with Oblimin Rotation

Factor Pattern Matrix[b]

Variables	1 Family /Negro	2 Socio-economic Status	3 East Indian	4 Socio-economic Status/ White	5 Family	6 Socio-economic Status/ Ethnic
1. Sex ratio					.9004	
2. Population aged 0-2 (%)	.8986					
3. Population aged 3-14 (%)	.9353					
4. Population aged 65+ (%)	.5217					-.5814
5. Adults never married (%)	.8827					
6. Person/household ratio			.3112		-.3336	
7. Live births, % per adult females at risk (aged 15-45)						
8. Adult females, % with 4 or more children					.6554	
9. Household-head sex ratio						
10. Adult males, % with secondary education		.3272				-.5705
11. Adult males, % with school certificate		.6967				
12. Adult males, % with university education				.8140		

Table 2.6 continued...

	1	2	3	4	5	6
13. Adult females, % with secondary education		.3760				-.5063
14. Adult females, % with school certificate		.6516				
15. Adult females, % in sales and clerical professions		.4767				
16. Adult females, % in professional occupations		.7621				
17. Negro population (%)	.9853					
18. White population (%)				.7031		
19. East Indian population (%)				.8671		
20. Chinese population (%)						-.7967
21. 'Mixed' population (%)						-.6729
22. Population of Anglican religion (%)	.9210					
23. Population of Roman Catholic religion (%)	.7110					
24. Population of Hindu and Muslim religions (%)	.4449		.6180			
25. Native Trinidadian population (%)	.8676					
26. Migrant population (%)	.6247	.3117				
27. Nonmover population (%)	.8460					
28. Density	.3439			-.3496		
29. Distance from CBD			.7154			
30. Area				.9267		

b--Loadings 0.30 are excluded. The principal loadings used to identify the dimension character are underlined.

TABLE 2.7

Oblique Structure Correlation Matrices*

1960

	Family Negro	Socio-economic Status White	East Indian	Family	Mobility	Socio-economic Status/ Ethnic
Family/Negro	1.000	.220	.346	.110	-.174	.218
Socioeconomic Status/White		1.000	.002	.097	.009	-.344
East Indian			1.000	.083	.095	.195
Family				1.000	-.030	-.376
Mobility					1.000	-.196
Socio-economic Status/Ethnic						1.000

1970

	Family/ Negro	Socio-economic Status	East Indian	White	Family	Socio-economic Status/ Ethnic
Family/Negro	1.000	.093	.152	-.137	.020	-.226
Socioeconomic Status		1.000	.190	.277	-.296	-.451
East Indian			1.000	.165	-.052	-.077
White				1.000	-.006	-.240
Family					1.000	.185
Socio-economic Status/Ethnic						1.000

* Factors are identified by their labels (see Tables 2.5 and 2.6).

family status variables; high numbers of infants (0.88) and young children (0.83), adults never married (0.66) and an aged population (0.41). The latter two associations are not surprizing given the family structure of lower-class West Indians (Rodman, 1971). The group appears mainly to be native Trinidadians (0.75), but there is little distinction between the groups' migration experiences. Apparently a nonmover population (0.81), there is representation of a migrant population (0.58). A very stong loading (0.94) identifies this group with the Anglican religion, but a relatively strong Roman Catholic association (0.61) suggests that religious differences within the Christian faith are not expressed in residential differentiation in Port of Spain.

(2) High socio-economic status, White group

This category is characterized by the association of a White ethnic identity (-0.74) with an idicator of high educational status; university education(-0.76) with low density residential living; area (-0.83) and with a male household head bias (-0.72). The latter distiniction separates this group from other West Indian groups in terms of a family status character.

(3) East Indian ethnic group

This group is a distinct ethnic-religious category of East Indians (0.94) who practice the Hindu and Muslim religions (0.85) and who live residentially separate from others in Port of Spain at some distance from the central business district (0.34).

(4) Family status category

A high positive loading (0.98) on the sex ratio associated with a high negative loading on the person per household ratio (-0.62) characterizes this category. A loading (0.39) of a family status measure of live births per adult females suggests there is residential separation between family life styles which is independent of ethnic differences as in factor 2, but consistent with a differentiation by sex in which higher proportions of females live in crowded quarters in larger families than do males.

(5) Mobility, inner city-periphery contrast

This category has a locational identity with negative loadings associated with an aged population (-0.43), with adults never married (-0.41), with a Chinese ethnic group (-0.38) and migrant status character (-0.36). It appears to represent the character of residential areas close to the central business district, distance (0.74).

(6) Middle socio-economic status ethnic group

A group of high positive loadings on educational and occupational variables identifies this category as a broad category of middle class status. The "mixed" Creole population (0.71) with a Chinese ethnic association (0.53) are associated strongly with the following variables: male and female secondary education (0.88), (0.82) respectively, male and female attainment of school certificate diplomas (0.70), (0.72) respectively, and female occupational categories in sales and clerical professions (0.90) and in professional occupations (0.63). Although it was anticipated that there would be more differentiation in the residential system according to these educational occupational strata, particularly because of social ecology theoretical arguments that have stressed the importance of female inclusion in the workforce as evidence of modernization (Timms, 1971), such a distinction is not evident in Port of Spain in 1960. Apparently there is a residential distinction that identifies a broad middle status group, but finer educational and occupational distinctions are not in evidence, except where they are reinforced by ethnic associations, as in Factors 2.1 and 2.2.

Extracting six dimensions from the 1970 comparable data set leads to the identification of several constructs which are similar to those of the 1960 factor solution.

(1) Family status, Negro group
This is characterized by the same high loadings that emerged in the 1960 factor solution, with one notable addition being a secondary loading of a religious variable, Hindu and Muslum (0.44). This would suggest that religious differences do not appear to be important discriminators of residential patterns in this Trinidadian city.

(2) Middle socio-economic status
This is a category that is distinguished by educational and occupational characteristics but has no ethnic association. It is similar to Factor 6 in the 1960 solution in terms of its high loading on male and female school certificate attainment (0.70), (0.65) respectively, and female occupational categories in sales and clerical professions (0.48), (0.76) respectively, representing a broad category with middle class status.

(3) East Indian ethnic group
The same high loadings that emerged in the 1960 factor solution identify this dimension; East Indian (0.87) with Hindu and Muslim (0.62) and with a stronger distance component (0.72) than previously.

(4) High socioeconomic status, White group
This group remains similar to its 1960 solution counterpart (Factor 2) in its loading associations; a white ethnic identity

(0.70), with an indication of university education (0.81), and a stronger area association (0.93).

(5) Family status category
The sex ratio bias that was evident in the residential system in the 1960 solution is maintained in 1970 as the same positive loading, sex ratio (0.90), associated with a negative loading on the person per household ratio (-0.34) occurs. However, there has been a change in family status association as the variable adult females with four or more children (0.66) is now linked to this dimension.

(6) Lower-middle socioeconomic status, ethnic group
An ethnic identity characterizes this group with high negative loadings on the Chinese (-0.80) and "mixed" Creole (-0.67), distinguishing this dimension from other socioeconomic status groups in 1970. Here, there is a distinction between educational attainment categories that was not evident in the 1960 factor solution. Variables depicting adult males with secondary education (-0.57) and females with secondary education (-0.51) are associated with this dimension, while higher levels of secondary education, i.e., school certificate attainment, in 1970 are associated with a "pure" socioeconomic status dimension (Factor 2).

Comparison of the inter-factor correlation matrices in Table 2.7, reveals a changing relationship between the dimensions that lends some support for the "convergence" proposition presented earlier. A primary family status/Negro dimension (Factor 1) has separated from other dimensions during the ten year period, with four of its correlations lower in 1970 than in 1960 (Table 2.7). Also, comparison of the factor patterns of 1960 (Table 2.5) and 1970 (Table 2.6) suggests that there has been a separation of a socioeconomic status construct from ethnic associations. In the 1960 factor pattern, the two socioeconomic status dimensions (Factors 2 and 6) have strong ethnic associations (Table 2.5). In the 1970 factor pattern, the second dimension extracted is a "pure" socioeconmic status construct, no longer associated with ethnicity (Table 2.6). Now (in 1970) and of lesser importance is the continued emergence of two socioeconomic status/ethnic dimensions (Factors 4 and 6) in the 1970 factor pattern (Table 2.6). Yet, the high inter-correlation (-.45) of this "pure" dimension, with one of these socioeconmic status/ethnic dimensions (Factor 6), suggests that separation is far from complete (Table 2.7). The continued merging of the white/socioeconomic status dimension with other ethnic dimensions between 1960 and 1970 (Table 2.7) is a continuation of the trend observed between 1931 and 1960 (Table 2.4). It seems that by 1970 there is the beginning of a reordering of the residential structure in Port of Spain, there is the emergence of a distinct socio-economic status dimension as a major latent construct of residential differentiation and there is a decline in importance of ethnic status differentiation.

If the longitudinal analysis of dimension separation has approached the question of change, the third phase is in part a verification procedure and in part an identification of stable relationships during the 1960-70 period. Canonical correlation analysis applied to comparable 1960 and 1970 data matrices identifies the continued existence in 1970 of relationships between variables already evident in the 1960 ecological structure.[4] Canonical correlation analysis extracts canonical variates according to relative degrees of serial (temporal) correlation between data sets. Thus we can gain an assessment of the most persistent elements in the ecological structures. In addition, the absence of variates will reinforce the findings of the 1960-70 longitudinal analysis concerning change in the ecological structures between 1960 and 1970.

The variables in each data set which have high loadings on the canonical variates are used to identify the character of "stable variates" in the two ecological structures (Table 2.8). Canonical variates are extracted until all variance between sets is accounted for, and and assessment of the relative importance of these stable variates is possible through the computation of redundancy measures (Briggs and Leonard, 1977).

An examination of the Proportioned Redundancy of the canonical variates (Tables 2.9) suggests that the first six variates of the 1960 set are significant contributors to explantion of the 1970 variance. These six appear to be the dominant persistent elements in Port of Spain's ecological structure throughout the 1960-70 decade. In order of importance they are:

(1) An East Indian variate explaining over 22 per cent of the 1970 variance.

(2) A family status/local born variate explaining 14 per cent of the 1970 variance.

(3/5) Two similar variates, which are socioeconmic status composites with ethnic associations explaining 16 per cent and 12 per cent respectively of the 1970 variance.

(4) A bi-polar variate representing ethnicity and socioeconomic

TABLE 2.8

Canonical Correlation Analysis: Port of Spain, 1960–70*

1960 SET

Variables	East Indian	Family/ Local Born	Socio-eco-nomic/ Ethnic	Bi-Polar Socioeco-nomic	Socio-eco-nomic/ Ethnic	Resi-dential Density	Proportional Variable Redundancy
	1	2	3	4	5	6	7
1. Sex ratio							.010
2. Population aged 0-2 (%)	.370	.542		-.514			.027
3. Population aged 3-14 (%)	.496	.674		-.539	.311		.039
4. Population aged 65+ (%)	.370	.426	.637		.613	.610	.044
5. Adults never married (%)	.315	.358				.346	.032
6. Person/household ratio	.376					-.565	.029
7. Live births,% per adult females at risk (aged 15-45)			-.339				.014
8. Adult females, % with 4 or more children			-.402	-.361	-.329		.020
9. Household-head sex ratio				.437			.016
10. Adult males, % with secondary education	.613	.542	.637		.529	.421	.050
11. Adult males, % with school certificate	.521	.403	.650		.539	.414	.046
12. Adult males, % with university education			.492	.497	.460		.028

Table 2.8 continued....

	1	2	3	4	5	6	7
13. Adult females, % with secondary education	.532	.501	.718		.621	.410	.053
14. Adult females, % with school certificate	.451	.363	.690		.591	.351	.048
15. Adult females, % in sales and clerical professions	.553	.508	.643		.516	.405	.046
16. Adult females, % in professional occupations	.387	.406	.640		.590	.361	.053
17. Negro population (%)		.349		-.390	.334		.015
18. White population (%)			.496	.495	.467		.027
19. East Indian population (%)	.746	.609	-.379	-.579	-.446		.058
20. Chinese population (%)	.301		.464		.337	.595	.033
21. 'Mixed'population (%)	.556	.619	.498		.406	.475	.045
22. Population of Anglican religion (%)	.318	.529	.359	-.409	.435		.035
23. Population of Roman Catholic religion (%)	.491	.593	.422	-.313	.425	.510	.040
24. Population of Hindu and Muslim religions (%)	.615	.557		-.508			.040
25. Native Trinidadian population (%)	.528	.705	.354	-.506	.373	.328	.044
26. Migrant population (%)	.424	.501	.388		.399	.482	.034
27. Nonmover population (%)	.530	.730	.320	-.558	.340		.044
28. Density	-.493	.438		-.670			.030

Table 2.8 continued....

1970 SET

Variables	East Indian	Family/ Local Born	Socio-eco-nomic/ Ethnic	Bio-Polar Socio-economic	Socio-eco-nomic/ Ethnic	Resi-dential Density	Proportional Variable Redundancy
	1	2	3	4	5	6	7
1. Sex ratio							.009
2. Population aged 0-2 (%)	.326	.421		-.384			.021
3. Population aged 3-14 (%)	.443	.609		-.504	.330		.032
4. Population aged 65+ (%)	.488	.602	.734		.697	.551	.065
5. Adults never married (%)	.432	.619		-.487	.361		.038
6. Person/household ratio			-.349			-.681	.031
7. Live births, % per adult females at risk (aged 15-45)							.010
8. Adult females, % with 4 or more children							.007
9. Household-head sex ratio							.006
10. Adult males, % with secondary education	.667	.685	.628		.540	.340	.063
11. Adult males, % with school certificate	.520	.431	.444		.335		.032
12. Adult males, % with university education			.460	.429	.427		.028
13. Adult females, % with secondary education	.620	.597	.669		.592	.312	.060

Table 2.8 continued....

	1	2	3	4	5	6	7
14. Adult females, % with school certificate	.477	.376	.360				.029
15. Adult females, % in sales and clerical professions	.411	.378	.419		.321		.024
16. Adult females, % in professional occupations	.305		.318				.040
17. Negro population (%)		.348		-.356	.426		.018
18. White population (%)			.474	.485	.449		.027
19. East Indian population (%)		.797	.604	-.353	-.558	-.402	.060
20. Chinese population (%)	.431	.314	.547		.406	.500	.051
21. 'Mixed'population (%)	.668	.673	.513		.395	.405	.051
22. Population of Anglican religion (%)		.493	.321	-.431	.425		.033
23. Population of Roman Catholic religion (%)	.579	.706	.499	-.372	.506	.412	.055
24. Population of Hindu and Muslim religions (%)	.682	.577		-.457			.040
25. Native Trinidadian population (%)	.559	.722	.376	-.507	.421		.050
26. Migrant population (%)	.439	.513	.424		.435	.357	.034
27. Nonmover population (%)	.542	.714	.331	-.535	.381		.048
28. Density	-.468	.460		-.689			.033
CANONICAL CORRELATIONS	0.97	0.55	-.69	0.75	0.59	0.68	

* The loadings used to identify the dimensions are underlined.

TABLE 2.9

Canonical Correlation Analysis Port of Spain, 1960-70
Proportioned Redundancy Reasures

CANONICAL VARIATES	1960 DATA SET Redundancy	Proportioned Redundancy	1970 DATA SET Redundancy	Proportioned Redundancy
1. East Indian	.176	.220	.197	.226
2. Family/Local born	.113	.141	.125	.144
3. Socioeconomic/Ethnic	.133	.166	.230	.264
4. Bi-Polar, Socioeconomic	.102	.127	.088	.101
5. Socioeconomic/Ethnic	.098	.122	.084	.096
6. Residential density	.082	.102	.055	.063
7.	.041	.051	.042	.048
8.	.015	.019	.013	.015
9.	.023	.029	.020	.023
10.	.012	.015	.009	.010
11-13.	.002	.002	.002	.002
	R = .801		R = .869	

status differentials; at one pole, the White, upper status group and at the other the Negro and East Indian low status groups. This explains 12 per cent of the 1970 variance.

(6) A residential density variate explaining 10 per cent of the 1970 variance (Tables 2.8 and 2.9).

What is clear from this canonical solution is confirmation that socioeconomic status and ethnicity differentials are the persistent elements in the ecological structure through the 1960-70 decade. Consistently, the East Indian group has remained residentially separated. Residential separation between White, upper status and Negro and East Indian low status groups appears to have been maintained. Socioeconomic status differentials also persist, although ethnic associations continue to qualify them. On the other hand, the absence of certain constructs may reflect non-persistent relationships. Particularly, the absence of a well-defined family status/Negro construct in the canonical solution (Table 2.8), when it has continually emerged as a primary dimension of residential differentiation in the longitudinal ecological analyses (Tables 2.3, 2.5 and 2.6), prompts the suggestion that change in family status differentials may have been important during the 1960-70 decade.

In sum, there appears considerable persistency in socioeconomic status and ethnicity status differentials. Certainly the residential structure of this city reflects a fragmentation in which the subordinate groups, the East Indians, and Negroes have remained residentially separated from each other as well as from the White upper status group. Yet, change has occurred in the selective separation of the family status and socioeconomic status dimensions during the 1931-1970 period. Whether this change has spatial regularities is the next question considered.

The Spatial Pattern of Port of Spain's Changing Residential Mosaic, 1960–70

In this second analytical procedure a classification of changing residential areas is derived, in order to establish whether there are any general spatial regularities in the alterations of the ecological structure during the 1960-70 decade. A comparative statics approach is used applying Image factor analysis to 1960 and 1970 conformable data sets. However, the Varimax (orthogonal) rotation rather than the Oblimin (oblique) rotation is used here. Whereas an oblique solution is more appropriate for the study of a process involving interdependencies between dimensions as in the first procedure, an orthogonal solution is more appropriate in this procedure, since independent factor scores can only be obtained from an orthogonal solution. These independent scores are required to represent each sub-areal unit's relative position on the main dimensions so that the classification of changing residential areas can be derived. As it turns out, while minor differences can be noted between oblique and orthogonal solutions,[5] with orthogonal rotation it is anticipated that the resultant spatial patterns may be simplified.

From the data sets for 1960 and 1970, six factors were extracted accounting for 70 per cent and 77 per cent respectively of the total variance in the original data. Factor robustness, invariance and reliability tests conducted on these Image factor analysis solutions established the general consistency of four major dimensions in both solutions (Conway and Haynes, 1977). These four dimensions summarized variations in family status (Factor 1), socioeconomic status (Factors 2 and 4) and ethnicity (Factor 3). Associated with Factors 1,2 and 4 are ethnic characteristics: a Negro association with the family status constuct (Factor 1), a Chinese and Mixed race association with the primary socioeconomic status construct (Factor 2), and a White association with the secondary socioeconomic status construct (Factor 4). The separate ethnic East Indian dimension is independent of any family status or socioeconomic status association (Table 2.10 and 2.11).

Although variable homogeneity tests of these four major

TABLE 2.10

1960 City of Port of Spain: Image Analysis with Varimax Rotation

Variables	Factor Loadings Matrix[c]					
	1	2	3	4	5	6
	Family/ Negro	Socioeconomic Status/Ethnic	East Indian	Socioeconomic Status/White	Mobility	Family
1. Sex ratio		.3250				.7016
2. Population aged 0–2 (%)	.8984					
3. Population aged 3–14 (%)	.9132					
4. Population aged 65+ (%)	.5971	–.5210			–.3548	
5. Adults never married (%)	.8519				–.3503	
6. Person/household ratio						–.6753
7. Live birth, % per adult females at risk (aged 15–45)		.4869				.4680
8. Adult females, % with 4 or more children		.5303				
9. Household-head sex ratio				.6829		
10. Adult males, % with secondary education	.3248	–.8609				
11. Adult males, % with school certificate		–.8205		.3640		
12. Adult males, % with university education		–.3527		.7576		
13. Adult females, % with secondary education		–.8668		.3026		

Table 2.10 continued....

	1	2	3	4	5	6
14. Adult females, % with school certificate		-.7928		.4885		
15. Adult females, % in sales and clerical professions		-.8841				
16. Adult females, % in professional occupations		-.6477				
17. Negro population (%)	.9557					
18. White population (%)		-.4448		.7584		
19. East Indian population (%)			.9086			
20. Chinese population (%)		-.5813			-.4510	
21. 'Mixed' population (%)	.4505	-.7271				
22. Population of Anglican religion (%)	.9046					
23. Population of Roman Catholic religion (%)	.8271	-.3828				
24. Population of Hindu and Muslim religions (%)	.3380		.8088			
25. Native Trinidadian population (%)	.9123	-.3154				
26. Migrant population	.8044	-.3014			-.3549	
27. Nonmover population (%)	.8921	-.3004				
28. Density				-.3455		
29. Distance from CBD			.5501		.5666	
30. Area				.7432		
Eigenvalues	10.79	6.80	2.15	1.72	1.03	.82

[c]Loadings 0.30 are excluded. The principal used to identify the dimension character are underlined.

TABLE 2.11

1970 City of Port of Spain: Image Analysis with Varimax Rotation

Variables	Factor Loadings Matrix[d] 1 Family/ Negro	2 Socioeconomic Status/White	3 East Indian	4 Socioeconomic Status/White	5 Family	6 Socioeconomic Status/Ethnic
1. Sex ratio					.6025	
2. Population aged 0-2 (%)	.8714					
3. Population aged 3-14 (%)	.9506					
4. Population aged 65+ (%)	.5864	.6719				
5. Adults never married (%)	.9157					
6. Persons/household ratio					-.3440	
7. Live births, % per adult females at risk (aged 15-45)		-.3981			.3388	
8. Adult females, % with 4 or more children					.5560	
9. Household-head sex ratio				.3499		
10. Adult males, % with secondary education	.4387	.7725				
11. Adult males, % with school certificate		.6480				.4154
12. Adult males, % with university education		.3198		.7955		
13. Adult females, % with secondary education	.3750	.7769				

Table 2.11 continued....

	1	2	3	4	5	6
14. Adult females, % with school certificate		.4419				.4519
15. Adult females, % in sales and clerical professions		.6224				.3757
16. Adult females, % in professional occupations						.5848
17. Negro population (%)	.9320					
18. White population (%)				.7620		
19. East Indian population (%)	.3195		.8439			
20. Chinese population (%)		.7282				
21. 'Mixed' population (%)	.3016	.8226				
22. Population of Anglican religion (%)	.9189					
23. Population of Roman Catholic religion (%)	.7869	.5454				
24. Population of Hindu and Muslim religion (%)	.5227		.6275			
25. Native Trinidadian population (%)	.9147	.3364				
26. Migrant population (%)	.6567	.4391				.4540
27. Nonmover population (%)	.9048					
28. Density	.3360			-.3964		
29. Distance from CBD			.7418			
30. Area				.8361		
Eigenvalues	10.86	5.39	1.86	1.45	0.88	0.64

[d] Loadings 0.30 are excluded. The principal loadings used to identify the dimension character are underlined.

dimensions prevented the use of the actual factor scores as objective measures of the changing character of residential areas during the 1960-70 decade, there was sufficient consistency between solutions to allow the interpretation of change along the two primary dimensions, the family status/Negro (Factor 1) and socioeconomic/ethnic dimensions (Factor 2), to be used as the basis of a classification of changing residential areas (Conway and Haynes, 1977).

Using the factor scores on the two primary orthogonal dimensions, a ranking technique places each sub-area in a sextile class for each time period, 1960 and 1970. The degree of shift of sub-areas between extreme classes then is used to identify stability and relative change during the decade. Two types of stable areas are categorized, Stable Cores and Zones of Relative Stability (Table 2.12). The criteria for Stable Core Areas is the lack of shift of sub-areas in the most extreme classes of the sextile ranking on both dimensions. Thus sub-areas ranked in sextile categories 1 and 2 or in 5 and 6 on the primary family status/Negro dimension for both periods are considered stable on that dimension. Similary sub-areas ranked in the upper and lower sextile categories on the second primary socioeconomic/ethnic dimension for both periods are stable core areas. Four types of stable core are possible: (i) high socioeconomic status, low family status, (ii) high socioeconomic status, high family status, (iii) low socioeconomic status, low family status, and (iv) low socioeconomic status, high family status (Table 2.12). The Zones of Relative Stability are classified as those sub-areas in which a shift of not more than one sextile category in either direction occurs between 1960 and 1970 on one dimension, while on the other dimension there is no shift out of an extreme class.

A similar procedure to identify Unstable Core areas and Zones of Relative Instability is followed (Table 2.12). Unstable Core areas are those sub-areas in which there is a shift of more than one sextile on the family status/Negro dimension, accompanised by a shift of one or more categories on the socioeconomic status/ethnic dimension. Zones of Relative Instability are identified if there is movement of more than one sextile category on the socioeconomic status/ethnic dimension accompanied by no movement or only one category shift between 1960 and 1970 on the family status/Negro dimension (Table 2.12). If movement is upward on the socioeconomic status dimension into sextile categories 1 or 2 in 1970, then this is classified as an upgrading or an 'appreciating' change. If movement on the socioeconomic status dimension is to sextile categories 5 or 6 in 1970, then this is classified as a downgrading of

TABLE 2.12

Classification of Changing Residential Areas

STABLE					UNSTABLE	
Stable Core Areas				Zones of Relative Stability	Unstable Core Areas	Zones of Relative Instablity
High SES Low Family	High SES High Family	Low SES Low Family	Low SES High Family	Stable on one dimension/ Movement of one sextile in either direction on another dimension	Movement of more than one sextile on both dimensions Movement of more than one sextile on one dimension/ One movement on the other	Movement of more than one sextile on SES dimension One or no movement on family dimension (Appreciating & Depreciating Zones)

'depreciating' change. In this way, Appreciating and Depreciating Zones of Relative Instability can be identified.

The spatial distribution of both stable and unstable areas in Port of Spain is illustrated in Figure 2.3. Examination of this distribution suggests there is little regularity in the pattern of changing residential areas. There is some evidence of an east/west division in which the western sector emerges as a more stable high socioeconomic status area, while the eastern sector contains nuclei of stable low socioeconomic status areas. Two depreciating zones of relative instability in the midst of the stable high socioeconomic status core in the west give that sector a distinctiveness, and only in the eastern sector and on the east side of the CBD are appreciating zones of relative instability in juxtaposition to unstable core areas (Fig. 2.3). In and around the CBD there has been a major change and an upgrading of the residential area character, which most likely reflects the impact of an urban renewal project conducted in that inner-city area since 1960.[6]

By and large, this classification indicates that most subareas have remained relatively unchanged throughout in 1960-70 decade. Stable Core areas and Zones of Relative Stability predominate throughout the city. It appears that there is considerable stability in the spatial pattern of residential differentiation, even though the previous analyses have identified elements of change and a beginning of the reordering of the residential structure by 1970.

Conclusions

This spatio-temporal factorial ecology of Port of Spain from 1931 to 1970 has identified long-term and short-term changes in the residential structure of a type of urban system seldom subjected to quantitative ecological analysis. The analytical procedures used here complemented each other and constitute a thorough examination of the aspatial and spatial characteristics of the city's residential system. Linked to theory on the relations between social structure and residential differentiation and utilizing an evolutionary model of residential areas differences, this spatiotemporal factorial ecology had a purpose that went beyond exploratory description. And, although the evidence was not sufficient to totally substantiate a general proposition which expected a reordering of Port of Spain's residential structure during neo-colonial transformations, there was some support forthcoming. It would appear that the pattern of the residential system of this primate city from 1931 to 1970 has continued to be dominated by ethnic and racial distinctions.

FIG. 2.3

Changing Residential Areas : 1960—70

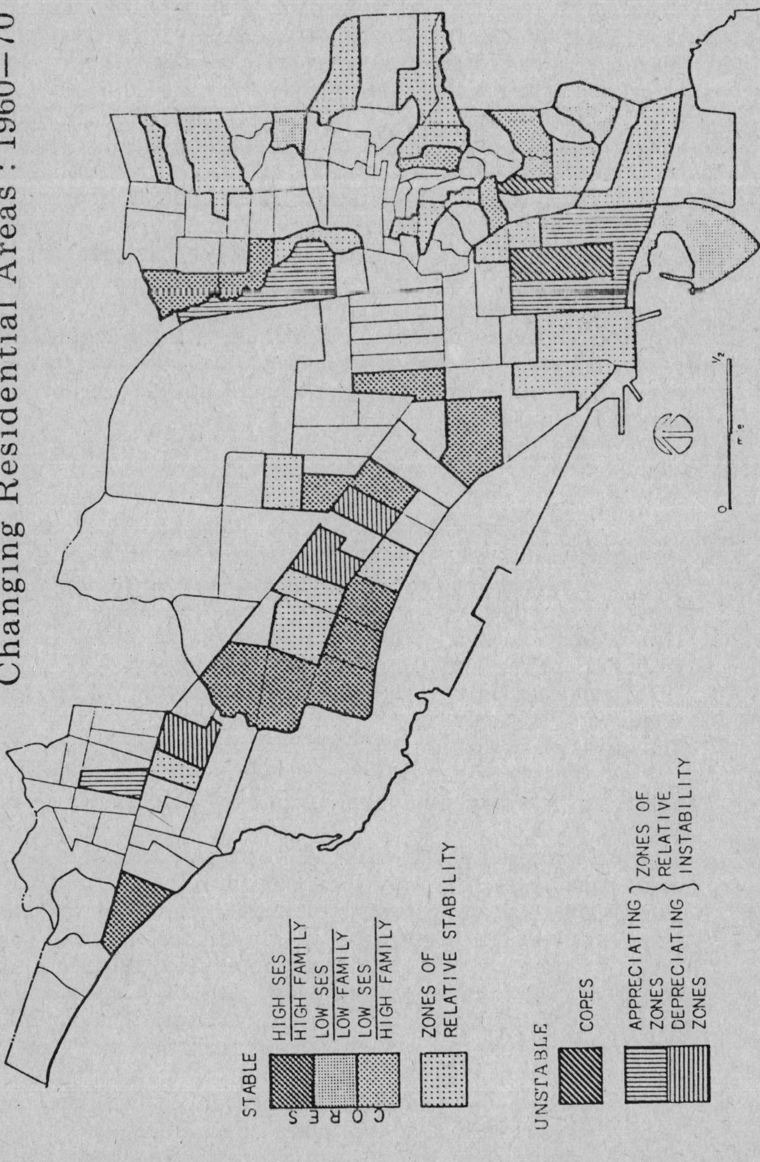

However, there is an indication that during the most-recent decade there has been some reordering.

The emergence of a "pure" socioeconomic status dimension in 1970 suggests there may be a beginning of a purely class-based reorganization of the residential structure, in this case possibly an emergent middle class, whose spatial manifestation is in land and housing consumption terms rather than in 'colonial' ethnic terms. Yet, alterations in the built environment and residential area reorganization appear to have been slow moving within the restricted city of Port of Spain.[7] Although the 1931 to 1970 period was one of political change, as Trinidad changed flags and moved from colonial to neo-colonial status, the response to these changes in terms of the city's internal residential structure has not been of any great magnitude. Perhaps most evident is the realization that the subordinate classes, the Negro and East Indian lower classes, continue to be residentially separated; a fragmentation along ethnic and migrant status differentials that evidences the persistence of the forces of the colonial era.

FOOTNOTES

1. It is not the main purpose of this study to argue about the strengths and weaknesses of existing urban ecological theory and to elaborate on the validity and appropriateness of the theoretical construct presented here. The theoretical position breifly summarized in this section is developed elsewhere (Briggs and Conway, 1975; Conway, 1976; Timms, 1971) and warrants a lengthy treatment that is beyond the scope of this paper.
2. The decision to retain six and five factor structures in this and all the other factor analysis solutions was based on the ability to identify and label individual factors (Conway and Haynes, 1973).
3. The Bajan category in the 1931 census indicates Port of Spain residents who were born in Barbados, a neighbouring West Indian island. Since the ethnic composition of this island territory is predominantly of African descent (77 per cent African, 5 per cent European and 18 per cent of Mixed origin), this Bajan category can be assumed to compose of residents who are racially Negro. The Trinidadian category in 1931 included persons classified both as Mixed and as Negro in the 1960 census.
4. Problems of data comparability prevent a similar analysis using the 1931 and 1960 restricted data matrices.
5. 1960 Varimax and Oblimin (Image factor analysis) solutions are virtually indistinguishable (Tables 2.5 and 2.11). In 1970,

a difference is the separation in the Oblimin solution of Factors 2 and 6 (Table 2.6), while in the 1970 Varimax solution, Factor 2 combines these dimensions (Table 2.11).

6. Evidence from aerial photography coverage of the city of Port of Spain in 1969 supports this suggestion (Conway, 1975).
7. This analysis has dealt solely with the internal reorganization of that part of the city of Port of Spain which has been continually urbanized since 1931. Omitted from this study is the reorganization which has occurred at the periphery beyond the municipal boundaries. Other work has addressed changes at the periphery utilizing different methodologies (Conway, 1976; Conway and Brown, 1980; Goodenough, 1975).

REFERENCES

Abu-Lughod, J. Testing the Theory of Social Area Analysis: The Case of Cairo. **American Sociological Review,** Vol. 34:(1969) 198-212.

———. The Legitimacy of Comparisons in Comparative Urban Studies: A Theoretical Position and Application to North African Cities. **Urban Affairs Quarterly,** Vol. 11:(1975) 13-35.

Ayub, M. and Cruickshank, E. The Political Economy of the Caribbean. **Finance and Development,** Vol. 14:(1977) 35-41.

Beckford, G.L. **Persistent Poverty.** New York: Oxford University Press, 1972.

Berry, B.J.L. Introduction: The Logic and Limitations of Comparative Factorial Ecology. **Economic Geography,** Vol. 47:(1971) 209-219.

Berry, B.J.L. and Rees, P.H. The Factorial Ecology of Calcutta. **American Journal of Sociology,** Vol. 74:(1969) 445-91.

Best, L. Outlines of a Model of Pure Plantation Economy. **Journal of Development Studies,** Vol. 17:(1968) 283-326.

Braithwaite, L. Social Stratification in Trinidad: A Preliminary Analysis. **Social and Economic Studies,** 2 and 3:(1953) 6-175.

Briggs, R. and Conway, D. The Evolution of Urban Ecological Structure: Theory and a Case Study, Port of Spain, Trinidad. **Social Science Quarterly,** Vol. 55:(1975) 871-88.

Briggs, R. and Leonard, W.A. Empirical Implications of Advances in Canonical Theory. **The Canadian Geographer,** Vol. 21:(1977) 133-47.

Brown, L.A. and Horton, F.E. Social Area Change: An Empirical Analysis. **Urban Studies,** Vol. 7:(1970) 271-88.

Chojnicki, Z. and Czyz, T. Some Problems in the Application of Factor Analysis in Geography. **Geographical Analysis,** Vol. 8:(1976) 416-27.

Clark, D., Davies, W.K.D. and Johnston, R.J. The Application of Factor Analysis in Human Geography. **The Statistician**, 23:(1974) 259-81.

Clarke, C.G. Pluralism and Stratification in San Fernando, Trinidad. **Social Patterns in Cities,** Institute of British Geographers, Special Publication No. 5:(1973) 53-70.

————. **Kingston, Jamaica: Urban Growth and Social Change, 1692-1962.** Berkeley: University of California Press, 1975.

Conway, D. The Use of Remote Sensing to Investigate Structural Change in Social Area Differentiation in Port of Spain, Trinidad (1975). In R.P. Momsen, Jr., (editor) **Geographical Analysis** for Development in Latin America and the Caribbean. Chapel Hill, N.C.: CLAG Publications.

Conway, D. **Residential Area Change and Residential Relocation in Port of Spain, Trinidad.** Unpublished Ph.D. dissertation: University of Texas at Austin (1976).

————. Urbanization and Dependent Capitalism: Whither Residential Differentiation? Presented in the Conference of Latin Americanist Geographers' Caribbean Urbanization Session. 75th Annual Meeting of the Association of American Geographers, Philadelphia (1979).

Conway, D. and Haynes, K.E. A Geographical Factor Analysis Model: Some Transatlantic Reflections. **Area,** Vol. 5:(1973) 106-9.

————. Advances in Comparative Ecological Analysis: Parsimony, Invariance and Homogeneity in Factor Analysis Solutions. **Environment and Planning A,** Vol. 9:(1977) 1143-56.

Conway, D. and Brown, J. Changing Relations of Intraurban Relocation and Structure: Low-Income Migrants in Latin America and the Caribbean. **Latin American Research Review,** (forthcoming) (1980).

Cross, M. **Urbanization and Urban Growth in the Caribbean.** Cambridge: Cambridge University Press (1979).

Davis, W.K.D. Alternative Factorial Solutions and Urban Social Structure: A Data Analysis Exploration of Calgary in 1971. **Canadian Geographer,** Vol. 22:(1978) 273-97.

Fanon, F. **The Wretched of the Earth.** Harmondsworth: Penguin Books (1967).

GIGGS, J.A. and MATHER, P.M. Factorial Ecology and Factor Invariance: An Investigation. **Economic Geography,** Vol. 51:(1975) 355-82.

Goodenough, S. Race, Status and Urban Ecology in Port of Spain, Trinidad. In C.G. Clarke (editor) **Caribbean Social Relations,** Liverpool: University of Liverpool, Centre for L.A. Studies, (1975) 8.

Greer-wootten, B. Changing Social Areas and the Intra-Urban Migration Process. **Revue de Geographie de Montreal,** Vol.

26:(1972) 271-92.

Harris, C.W. **Problems in Measuring Change.** Madison, WI: University of Wisconsin Press (1963).

Harvey, D. **Social Justice and the City.** London: Edward Arnold (1973).

Hawley, A.H. **Urban Society: An Ecological Approach.** New York: Ronald Press (1971).

Haynes, K.E. Spatial Change in Urban Structures: Alternative Approaches to Ecological Dynamics. **Economic Geography,** Vol. 47:(1971) 324-35.

Hunter, A.A. The Ecology of Chicago: Persistence and Change, 1930-60. **American Journal of Sociology,** Vol. 77:(1971) 425-44.

Johnston, R.J. Some Limitations of Factorial Ecologies and Social Area Analysis. **Economic Geography,** Vol. 47:(1971) 314-23.

————. Social Area Change in Melbourne, 1961-66. **Australian Geographical Studies,** Vol. 11:(1973a) 79-98.

————. Possible Extensions to the Factorial Ecology Method: A Note. **Environment and Planning,** 5:(1973b) 719-34.

————. Residential Area Characteristics: Research Methods for Identifying Urban Sub-Areas--Social Area Analysis and Factorial Ecology. In D.T. Herbert and Johnston, R.J. (editors) **Social Areas in Cities,** Volume 1, **Spatial Processes and Form,** Chichester, Sussex: John Wiley, (1976) 193-235.

————. On the Characterization of Urban Social Areas. **Tijdschrift voor Economische en Sociale Geografie,** 70:(1979) 232-38.

Klaff, V.Z. Ethnic Segregation in Urban Israel. **Demography,** 10:(1973) 161-184.

Lo, C.P. Changes in the Ecological Structure of Hong Kong, 1960-71: A Comparative Analysis. **Environment and Planning A,** Vol. 7:(1975) 941-63.

Lojkine, J. Contribution to a Marxist Theory of Capitalist Urbanization. In C.G. Pickvance (editor) **Urban Sociology: Critical Essays,** London: Methuen, (1972) 119-146.

London, B. and Flanagan, W.G. Comparative Urban Ecology: A Summary of the Field. In J. Walton and L.H. Masotti (editors) **The City in Comparative Perspective,** New York: Halsted Press (1976) 41-66.

Loweenrhal, D. **West Indian Societies,** London: Oxford University Press (1972).

Masotti, L.H. and Walton, J. Comparative Urban Research: The Logic of Comparisons and the Nature of Urbanism. In J. Walton and L.H. Masotti (editors) **The City in Comparative Perspective,** New York: Halsted Press, (1976) 1-15.

McElrath, D.C. Societal Scale and Social Differentiation: Accra, Ghana. In S. Greer (editor) **The New Urbanization,** New York: St. Martin's Press (1968).

Murdie R.A. **Factorial Ecology of Metropolitan Toronto, 1951-61**, Research Publication 116, Department of Geography, University of Chicago (1969).

Newton, P.W. and Johnston, R.J. Residential Area Characteristics and Residential Area Homogeneity: Further Thoughts on Extensions to the Factorial Ecology Method. **Environment and Planning A**, Vol. 8 (1976) 543-52.

Oechsli, F.W. and Kirk, D. Modernization and the Demographic Transition in Latin America and the Caribbean. **Economic Development and Cultural Change**, Vol. 23 (1975) 391-419.

Perle, E.D. Scale Changes and Impacts on Factorial Ecology Structures. **Environment and Planning A**, Vol. 9 (1977) 549-58.

————. Target Rotation and Normative Policy Analysis in Urban Ecology. **Environment and Planning A**, Vol. 10:(1978) 275-85.

Rampersad, F. **Growth and Structural Change in the Economy of Trinidad and Tobago 1951-1961**. Institute of Social and Economic Research: University of the West Indies, Jamaica.

Ramsaran, R. Integration and Underdevelopment in the Commonwealth Caribbean. **Intereconomics**, (1977) 197-200.

Rodman, H. **Lower Class Families: The Culture of Poverty in Negro Trinidad**. London: Oxford University Press (1971).

Ryan, S. **Race and Nationalism in Trinidad and Tobago**. Toronto: University of Toronto Press (1972).

Schilderinck, J.H.K. **Factor Analysis Applied to Developed and Developing Countries**. Rotterdam: Rotterdam University Press (1970).

Schwirian, K.P. **Comparative Urban Structure**. Lexington, Mass: D.C. Heath.

Schwirian, K.P. and SMITH, R.K. Primacy, Modernization and Urban Structure: The Ecology of Puerto Rican Cities. In K.P. Schwirian (editor), **Comparative Urban Structure**. Lexington, Mass: D.C. Heath, (1974) 324-38.

Semple, R.K., Youngmann, C.E. and ZELLER, R.E. Economic Regionalization and Information Theory: An Ohio Example. Columbus, Ohio: Ohio State University, Department of Geography Discussion Paper, 28 (1972).

Shevky, E. and Bell, W. **Social Area Analysis**. Stanford: Stanford University Press (1955).

Smith, M.G. **The Plural Society in the British West Indies**. Berkeley: University of California Press (1965).

Sweetzer, F. **Patterns of Change in the Social Ecology of Metropolitan Boston: 1950-1960**. Boston: Massachusetts Department of Mental Health (1962).

Timms, D.W.G. Modernization and the Factorial Ecology of the Cook Islands, Brisbane and Auckland. **Australian and New Zealand Journal of Sociology**, Vol. 6 (1970) 139-49.

________. **The Urban Mosaic.** Cambridge University Press (1971).

Walton, J. Political Economy of World Urban Systems: Directions for Comparative Research. In J. Walton and L.H. Masotti (editors) **The City in Comparative Perspective,** New York: Halsted Press (1976) 301-13.

Wilson, G. and WILSON, M. **The Analysis of Social Change,** Cambridge: Cambridge University Press (1945).

THREE

Christina R. Victor

The Urban Dimensions of Greater Nottingham, England: 1971

INTRODUCTION

Examination of the relevant literature reveals that the vast majority of British factorial ecologies have used either 1961 (Herbert, 1970) or 1966 (Evans, 1973) census data as their information source. Few researchers have explored the potential of the most recent enumeration conducted in 1971. Largely this reflects the decline in popularity which factor analysis has recently encountered in the geographical world. In this paper a factorial ecology of the Greater Nottingham urban area, using 1971 census information, is described. The particular objectives are fourfold. Firstly the nature of socio-spatial segregation in the study area is described. The results are then compared with those published by earlier British workers and their degree of generality established. Comparative reference is also made to the classic triple axes model of differentiation proposed by the Social Area Analysts. In the final part of the report a number of conclusions concerning the structure of segregation in British cities are suggested.

Methodology

Having outlined the objectives of the study and its theoretical background, the methodological basis of the investigation is now described. Included within this section is a brief description of the study area, the technical procedures used and the choice of indices for inclusion in the analysis.

The Study Area

For this exercise the study area was defined as consisting ot the pre-1974 local government reorganisation areas of Nottingham County Borough and the five contiguous Urban Districts of Arnold, Beeston and Stapleford, Carlton, Hucknall and West Bridgford. Greater Nottingham, the amalgam of these administrative units, was considered to form a functionally interdependent urban unit (Victor, 1979). This contrasts with the type of study area traditionally used by British urban geographers. Study areas have usually been defined in terms of an adminsitrative definition of the city (Davidson, 1975). The major exception to this broad generalisation is the work of Davies and Lewis (1973) in Leicester.

In 1971 a population of 497,859 was enumerated in Greater Nottingham. Of this 300,675 were resident in the County Borough with the remainder 197, 184, resident in the suburban ring (i.e., the five contiguous Urban Districts). Victor (1979) has documented the many social disparities which exist between the city and the surrounding suburbs. In population terms the study area represented the twelfth largest urban area in Great Britain and functioned as the capital of the East Midland region.

Technical and Analytical Procedures

Two basic types of multivariate statistical procedure have been used in factorial ecologies. British workers have tended to favour Priniciple Components Analysis (P.C.A.) whilst North American researchers have favoured true Factor Analysis. Different statistical assumptions are characteristic of these two techniques (Mather, 1975). Some debate has taken place in the literature concerning the "correctness" of rotating components to a position of simple structure. As a rotation procedure was envisaged in this research strategy, as a means of defining specific and unambiguous dimensions of differentiation, a factor analysis technique senso stricto was employed.

A number of different types of factor analysis are available to the social scientist (Mather, 1975). It has been suggested that the results of any factorial ecology are directly dependent upon the specific technique employed (Rees, 1971). Recently, however, Giggs and Mather (1975) have disputed this suggestion. These two authors have specifically investigated the influence of factoring technique upon the resultant factor structure of Nottingham County Borough. This technique comparative survey revealed the similarity, rather than differences, existing between the five major types of factoring procedure included in the study. Consequently for this investigation the familiar priniciple axes mode of analysis was used in preference to some of the newer and less well-known factoring routines now available.

Two basic types of rotation procedure are available to the factorial ecologist. Traditionally British factorial investigation have employed an orthogonal solution (Robson, 1969), usually the Varimax solution developed by Kaiser (1958). This type of solution ensures that the resultant factor axes remain independent (i.e., uncorrelated) from each other. It has, however, been proposed that this specification is an unrealistic representation of the social world (Bell, 1955; Giggs and Mather, 1975). Thus oblique rotation procedures, which allow for the possibility that the factor axes may be inter-correlated, have been developed. For this work a less deterministic oblique rotation procedure was favoured. The direct oblimin method, with the data value set at zero, was the specific technique employed.

In any factorial ecology, the selection of the indices is a crucial decision which has to be taken by the researcher. Any biases or deficiencies present in the original data matrix will obviously be reflected in the resultant factor structures. By grouping the indices into "like-type" groups, as in Table 3.1, any over or under representations are made explicit. Surprisingly few factorial ecologists have used this simple technique to check the spread of their information. Within the conceptual limitations imposed by the use of census data as an information source in urban studies (Smith, 1973; Johnston, 1978). Table 3.1 illustrates that the measures included are broad representations of the major components of British urban social structure.

TABLE 3.1

Variables Employed in the Multi-variate Analysis Social Structure

Population structure

% Total population aged 0 to 14 years
% Total population aged 20 to 44 years
% Total population aged 45 to 59 years
% Total population aged 65+ years
% Total population over 15 years unmarried (single, divorced or widowed)

Family status

Fertility ratio - number of children aged 0 to 4 years per 1,000 women aged 15 to 44.

% Total households with 1 or more children aged 0 to 4

years
- % Total households with 1 or more children aged 5 to 14 years
- % Total households single person (not pensioner)
- % Total households with 1 or more pensioners

Ethnic status
- % Total population born in the New Commonwealth (Africa, America, India, Pakistan)
- % Total population born in Ireland, Europe and the Old Commonwealth (Canada, Australia and New Zealand)
- % Total population with 1 or both parents born in the New Commonwealth

Economic participation
- % Total population economically active
- % Economically active population unemployed
- % Economically active population female
- % Economically active population with OND, HND and degree qualifications

Migration
- % Total population 1-year intra-Local Authority movers
- % Total population 1-year inter-Local Authority movers
- % Total population 5-year intra-Local Authority movers
- % Total population 5-year inter-Local Authority movers

Socio-economic status
- % Total households in Class I (by SEG of household head)
- % Total households in Class II
- % Total households in Class IV
- % Total households in Class V

Affluence
- % Total households with 1 car
- % Total households with 2 or more cars

House tenure
- % Households owner-occupiers
- % Households Local Authority renters
- % Households private rented unfurnished
- % Households private rented furnished

Dwelling condition
- % Total households sharing a dwelling
- % Total households overcrowded (1.5+ persons per room)
- % Total households who share/lack hot water
- % Total households who share/lack fixed bath or shower

% Total households who share/lack inside WC

Dwelling size

% Total dwellings with 1 to 3 rooms
% Total dwellings with 7+ rooms

The Empirical Results

In the following section the empirical results of the survey are described. These findings are summarised in two parts. The first pertains to the standard primary order factor analysis. In the second sub-section a more exploratory higher order factor solution is described. In the final section of the paper these empirical results are compared with previous British studies and the theoretical postulates of urban social structure derived in the North American context.

The Primary Analysis

For the primary analysis the 38 variables were computed for the 983 enumeration districts defined for the collection of the 1971 census. Application of the statistical procedures described above produced eight factors with eigenvalues of one or over (Table 3.2). Almost three-quarters of the variance present in the initial data set was represented by these composite dimensions. Rotation of the factors by the direct oblimin method produced two factor loading matrices - namely the factor structure and factor pattern matrices. Some debate has centred upon which is the "correct" matrix to interpret. For example, Timms (1970) used the factor structure matrix, whilst in contrast Harman (1970) and Giggs (1973) employed the factor pattern matrix. However, Rummel (1970), considered that the pattern matrix was the more relevant of the two. Hence the pattern matrix formed the bais for the interpretation of the factorial ecology of Greater Nottingham (Table 3.3). In interpreting the nature of each dimension, only those loading significant at the 99.9 per cent level were used (i.e. 0.3 and over).

Limitations of space constrain the discussion to only a brief description of the nature of the individual dimensions. A fuller exposition is available elsewhere (Victor, 1979). Summarised the eight factors represented:

(I) Socio-economic status—this is an unambiguous measure characterised by positive loadings on the high status (Class I) and affluence (car and house ownership) measures.

TABLE 3.2

Greater Nottingham: Factor Explanation: The First Order Solution

Factor	Eigenvalue	Variance (%)	Cumulative Variance (%)
1	8.01	23.9	23.9
2	5.40	14.9	38.8
3	4.10	9.5	48.3
4	3.00	8.3	56.6
5	2.50	6.6	63.2
6	1.80	4.7	67.9
7	1.50	4.3	72.2
8	1.07	2.7	74.9

(II) House tenure—households renting accommodation from the Local Authority (i.e., public rented housing) are contrasted with home owners or those who rent from a private landlord. Position in the housing market is clearly identified as an important aspect of social differentiation.

(III) Urbanism—a negative orientation is characteristic of this factor. Significant loadings with the shared dwellings, small households, ethnic minority and private rented housing variables identify a 'classic' urbanism dimension.

(IV) Life-cycle stage—the familistic life style may be subdivided into a number of separate stages of development. Factor four contrasts the post child phase, identified by positive loading indices, with the child centred phases.

(V) Economic participation—with positive loadings on the variables measuring the proportion of persons employed and women employed, this is a clear measure of economic participation variations.

(VI) Ethnicity—ethnic orgins, intra-urban mobility and unemployment are the three sets of social attributes summarised by factor six.

(VII) Female employment—this is a highly specific factor characterised by a single diagnositc loading - the proportion of adult females gainfully employed.

(VIII) Tenure—life-cycle—again this is a highly specific

TABLE 3.3

Greater Nottingham: Factorial Ecology: Factor Pattern Matrix

	FACTOR							
	1	2	3	4	5	6	7	8
% Population 0-14	0.17	0.36	0.61	-0.82	0.23	0.13	-0.06	-0.07
% Population 20-44	-0.04	-0.05	-0.65	-0.58	0.51	0.07	-0.05	-0.02
% Population 45-59	-0.004	0.03	0.11	-0.28	0.39	-0.08	-0.10	0.21
% Population 65+	-0.02	0.04	-0.06	0.61	0.26	-0.20	-0.23	0.02
% Adults single	-0.11	-0.01	-0.36	-0.05	0.41	0.01	0.16	-0.15
Fertility ratio	0.01	-0.06	-0.06	-0.36	0.07	0.01	-0.03	-0.02
% Households with children 0-5	0.23	0.07	0.19	-0.51	0.18	0.17	-0.01	0.41
% Households with children 5-14	0.19	0.12	0.07	-0.42	0.09	0.15	-0.08	0.28
% 1 person households	-0.26	0.06	-0.53	0.26	-0.05	-0.21	-0.17	0.02
% Pensioner households	-0.04	0.13	0.07	0.49	-0.01	-0.15	0.21	0.06
% Foreign born coloured	-0.53	-0.03	-0.52	0.04	0.06	0.65	0.03	0.03
% Foreign born white	-0.14	0.15	0.01	0.10	0.40	-0.01	-0.04	0.19
% Population parents coloured	0.23	0.19	-0.31	0.14	0.07	0.59	0.002	0.18
% Owner occupiers	0.57	-0.92	-0.06	-0.06	0.07	0.09	-0.15	0.02
% Local Authority renters	-0.06	0.54	-0.12	0.01	-0.09	-0.003	-0.21	0.32
% Private unfurnished	-0.02	-0.88	-0.4	0.09	0.02	0.08	-0.05	0.01

Table 3.3 continued.......

	1	2	3	4	5	6	7	8
% Private furnished	-0.01	-0.4	-0.41	-0.02	0.20	0.11	0.04	0.03
% Shared dwellings	0.03	0.04	-0.71	-0.05	-0.002	0.07	0.04	0.03
% Overcrowded	-0.13	0.23	-0.4	-0.01	0.09	0.22	-0.28	-0.05
% Share/lack hot water	-0.19	0.05	0.22	-0.05	-0.07	0.65	0.11	-0.07
% Share/lack fixed bath	-0.26	0.04	0.12	-0.02	-0.06	0.03	0.08	-0.04
% Share/lack inside WC	-0.27	0.09	-0.21	0.03	-0.01	0.03	0.02	-0.08
% 1-3 rooms	0.17	0.10	-0.52	0.05	-0.009	-0.25	-0.03	0.08
% 7+ rooms	0.05	-0.29	-0.75	0.16	-0.11	0.10	0.04	-0.25
% 1year intra-Local Authority movers	0.04	0.02	0.02	-0.04	0.02	0.37	0.10	-0.03
% 1 year inter-Local Authority movers	-0.07	-0.31	0.11	-0.01	0.005	0.03	-0.006	-0.01
% 5 years intra-Local Authority movers	0.07	0.02	0.008	-0.04	-0.004	0.47	-0.02	-0.04
% 5 years inter-Local Authority movers	-0.09	-0.61	0.02	-0.06	0.02	-0.02	-0.01	0.05
% Class I	0.52	0.09	0.07	0.05	-0.08	-0.09	0.05	-0.13
% Class II	0.27	0.01	-0.27	-0.06	0.05	-0.08	-0.09	0.05
% Class IV	-0.32	-0.01	0.04	-0.008	0.10	-0.07	-0.09	-0.14
% Class V	-0.61	-0.23	0.05	0.11	0.003	-0.05	0.06	-0.17
% Economically active	0.05	-0.02	0.01	0.10	0.6	0.04	-0.11	-0.01
% Unemployed	-0.08	0.04	0.07	0.03	0.11	0.65	-0.05	-0.19
% Employed female	-0.03	-0.04	0.06	-0.08	0.57	0.10	0.71	0.07
% Employed qualified	0.04	0.05	0.01	0.02	-0.02	0.04	0.17	0.23
% 1 car	0.69	-0.26	-0.17	-0.23	0.24	0.12	-0.22	-0.07
% 2 cars	0.71	0.21	-0.17	0.02	0.03	0.17	-0.22	0.06

factor in structure which is described by loadings with the child bearing and public housing variables.

Higher Order Factorial Ecology

By employing an oblique rotation procedure it is possible to examine the degree of inter-dependence existing between the eight primary factors. This data, which is expressed in terms of a correlation matrix, is shown in Table 3.4. It is evident from this that the eight primary axes are inter-correlated although to varying degrees. For example, status (factor I) exhibits a very strong negative relationship, -0.70, with urbanism (factor III). Whilst the majority of the linkages identified are not particularly large in magnitude, the employment of an orthogonal rotation procedure would clearly have been unduly deterministic in this example.

Demostration of these relationships between the primary factors suggested the presence of further 'generalizations' underlying the basic factor strkucture. Several workers have suggested that, by refactoring the inter-factor correlation matrix, it is possible to explicitly articulate these generalizing dimensions (Davies and Lewis, 1973; Giggs and Mather, 1975). Application of the statistical techniques described earlier produced three second order factors with eigenvalues of one or over (Table 3.5). The resultant factor pattern matrix is listed as Table 3.6.

Interpretation of this table resulted in the description of the three second order factors as representing:

(I) **Life-cycle status**—second order factor I illustrates positive relationships with primary factors IV (life-cycle status) and VIII (tenurelife-cycle) and negative linkages with the two economic participation indices (factors VII and V). Therefore, this first second order generalization is an unambiguous measure of variations in life-cycle status and economic participation in the study area.

(II) **Socio-economic status**—the second higher order factor is described, in structural terms, by three primary factors. These are factors I (status), II (tenure mobility) and III (urbanism). In structural terms this may be considered as a socio-economic status dimension with important house tenure and family status overtones.

(III) **Tenure - life-cycle status**—Statistically significant loadings with factor II (tenure), IV (ethnicity), and VII(tenure-

TABLE 3.4

First Order Factors : The Inter-Correlation Matrix

FACTOR	LABEL	1	2	3	4	5	6	7	8
1	STATUS								
2	TENURE-MOBILITY	-0.32							
3	URBANISM	-0.70	0.16						
4	LIFE CYCLE STAGE	-0.20	-0.09	-0.21					
5	ECONOMIC PARTICIPATION	0.12	-0.15	-0.29	-0.23				
6	ETHNICITY UNEMPLOYMENT	-0.14	-0.68	0.16	0.23	0.21			
7	EMPLOYED FEMALES	-0.17	-0.45	-0.10	-0.26	0.27	-0.05		
8	TENURE-LIFE CYCLE	0.10	0.29	-0.23	0.03	-0.28	0.23	-0.17	

TABLE 3.5

Second Order Factor Analysis : Dimensions of Variation

Factor	Eigenvalue	Variance (%)	Cumulative Variance (%)
1	2.94	36.8	36.8
2	2.10	26.3	63.1
3	1.20	15.1	78.2

TABLE 3.6

Second Order Factorial Ecology : Factor Pattern Matrix

First order factors	Second order factors 1	2	3
1	-0.01	-0.91	-0.12
2	0.17	-0.31	0.89
3	0.03	0.91	-0.06
4	0.40	0.10	0.02
5	-0.90	-0.13	0.25
6	0.12	0.09	-0.87
7	-0.70	0.12	-0.13
8	0.43	-0.16	0.37

life-cycle) are diagnostic of second order factor III. This dimension links young families with the public rented sector of the housing market and is contrasted with the private rented sector which is typified by ethnic minorities and intra-urban mobility.

Discussion

The results of these two factorial investigations are now discussed with specific reference to the major postulates of urban theory. The results are also compared with those reported by previous British workers and their degree of generality established.

In theoretical terms urban studies remains an underdeveloped field of study. Of the theoretical postulates which have been derived, two basic types may be recognised. The first aspect of urban theory pertains to the pattern of residential segregation as typified by the work of Burgess (1924) and Hoyt (1939). Shevky and Bell's (1955) writings considered the second major facet of urban theory, the structure of residential segregation. Only recently with the development of factorial ecology have the two been linked together.

As this particular paper is concerned with the nature, rather than pattern of segregation, these results were compared only with the work of Shevky and Bell (1955). The work of these two authors in general, and the social area analysts in particular, are well known. Very briefly summarised, therefore, they considered that social rank, urbanisation and ethnicity constituted the basic facets of urban social segregation.

Examination of the eight primary factors reveals that status (factor I), housing tenure (factors II and VIII), urbanism life-cycle status (factors III, IV), ethnicity (factor VI)and economic participation (factors V and VII) were the dimensions of segregation identified in Greater Nottingham. There is only a limited degree of correspondence between these dimensions and the social area axes. Whilst social and family status are important in Greater Nottingham, ethnic status is of much more limited importance. It is apparent that the totally different structure of the housing and employment markets in Britain and North America is reflected in the disparity between the Greater Nottingham results and social area postulates.

The results of the more generalized higher order investigation reveal a similar disparity. Status and life-cycle status measures are identificable, but not an ethnicity dimension. Indeed the results suggest that the status-urbanism dimensions may not be independent of each other as Shevky and Bell (1955) proposed. At both primary and higher order factor analysis levels there is, therefore, obvious disparity between the writings of the social area analysts and the empirical Nottingham results.

In seeking to explain this disparity it is necessary to consider both the temporal and cultural disjunction exhibited between Greater Nottingham and the North American context in which the theory was formulated. In addition it is necessary to consider the basic differences which are displayed between the city in North America and the British city. Firstly the historical legacy of the past is of much greater importance in the British city and may directly influence modern socio-spatial patterns. For example, much of the current housing problem in Nottingham C.B. may be directly attributed to the scale and pattern of 19th century expansion and development (Victor,

1979). The scale of ethnic minorities in British cities does not compare with the size and problem of immigrant and racial groups in North American cities.

A third important difference between the British and North American city is the scale of Local Government intervention. With a large percentage of Local Authority built and subsidised dwellings and a high degree of town planning the British city contrasts strongly with its American counterpart. In addition to this, however, British cities receive a larger degree of central government funding than their North American equivalents. These specific urban differences in conjunction with the general social and cultural differences exhibited between Britain and North America, largely explain the disparity illustrated in urban structure. For its theoretical constructs, therefore, British urban geographers should look to their own cultural context.

Comparision with previous British studies

By comparing the results of the Nottingham study with those produced by previous British studies, it is possible to identify generalizations concerning urban social structure. The establishment of such generalizations is the first step to forming a theory of British urban social structure. Two comparative aspects are discussed in some depth. The first relates to the factoring methodology, whilst the second pertains to the resultant solution. Traditionally, British researchers have employed a unique set of variables in their factor studies (Table 3.7). A wide variation in the number of variables used is evident ranging from twenty-six (Herbert, 1970) to ninety-six (Edwards, 1974). Differences in the nature of the employed indices are also apparent. Thus Robson (1969) failed to include any measures of ethnicity in his study of Sunderland, whilst Thomas (1968) excluded residential mobility measures from his research scheme. Whilst the breadth of social structure measured shows considerable inter-study variation, most researchers have strongly emphasised housing tenure and quality in their investigations.

It is, however, a contentious issue as to whether a uniform optimal set of indices should be defined to provide the empirical basis of future factorial investigations. Herbert (1970) comprehensively demonstrated that the variables proposed for national use by the Centre for Urban Studies (Norman, 1969) were of limited usefulness in studying socio-spatial differentiation in South Wales. Davies (1975) has expanded the debate in this field. He has argued that the importance of the dimensions of differentiation and their definition would vary with the place

TABLE 3.7

Variables Input for Selected British Factorial Ecologies

STUDY	TOTAL NO. OF VARIABLES	SOCIO-ECONOMIC	DEMO-GRAPHIC	ETHNIC	HOUSE-HOLD	MOBILITY	HOUSING CONDITION	TENURE	CTHER HOUSING	TOTAL HOUSING	OTHER
CARDIFF and SWANSEA	26	3	6	7	1	1	4	4	-	9 (34.6%)	17
SUNDERLAND	30	8	6	-	6	-	5	4	1	16 (53.3%)	14
LEICESTER	55	24	8	4	3	5	4	3	2	12 (21.8%)	43
CARDIFF, SWANSEA NEWPORT	40	15	7	1	3	2	4	3	5	15 (37.5%)	25
NOTTINGHAM	31	13	2	2	1	-	5	5	2	13 (41.9%)	18
GREATER NOTTINGHAM	38	11	6	3	3	4	6	4	2	14 (37.8%)	23
CARDIFF	33	5	11	3	-	4	3	4	3	10 (30.3%)	23
HULL	35	10	9	1	5	1	4	4	1	14 (40.0%)	21

in the urban hierarchy of the study area. Hence several idealized data sets would be required and suggests that the search for one specific data set is misleading.

Table 3.8 reveals, however, that variation in British factorial studies is not restricted to matters of data input; differences in data source, level of spatial generalization and technique are all obvious. Each of these variations may well crucially influence the degree to which the results of individual studies may be compared. In order to campare the empirical results of these studies, the general structure of all factors/components with eigenvalues in excess of unity, are discussed in terms of their labels. Although Palm and Caruso (1972) are highly sceptical of the practice of factor/component labelling, it is now established practice and provides a simple method of comparing structures.

Of the selected eleven studies, eight identified socio-economic status as the leading dimension of urban social segregation. Included within this group was the present Nottingham study. A second general dimension appears to be housing tenure—a facet which was not identified in only two study areas, family and life-cycle status dimensions also seem to be general to the majority of studies. However, it would appear that a housing quality dimension is only of particular importance in studies which used pre-1971 data. This suggests that housing quality may have declined in importance as a differentiator of the British urban population between 1961 and 1971. One possible explanation for this observation may be that the extreme housing quality differentials, characteristic of British urban areas in 1960, may have been eroded by the comprehensive urban renewal programmes of the decade 1960-1970. Evidence to support this hypothesis was certainly identifiable in the Greater Nottingham urban area (Victor, 1980). The remaining structures show little generality and appear to be specific to the individual study areas. Significantly, ethnic status measures are included within this latter group and do not seem to be a general feature of British urban social segregation.

Examination of the comparative results survey reveals that it is possible to identify two types of factor structure. These consist of the general dimensions of urban social structure and the place specific elements. Facets such as status, family status—life-cycle stage and house tenure constitute the aspects of urban social structure common to most British urban areas. In addition, within any study area it is also possible to identify place specific elements in the urban structure. Theories which seem to explain the nature of British urban social segregation

TABLE 3.8

Selected British Factorial Ecologies : Results Summary

STUDY AREA	AUTHOR	DATA BASE YEAR	NUMBER OF AREAL UNITS	FACTORING METHOD	DIMEN-SION ONE	DIMEN-SION TWO	DIMEN-SION THREE	DIMEN-SION FOUR	DIMEN-SION FIVE	DIMEN-SION SIX	DIMEN-SION SEVEN	DIMEN-SION EIGHT
1	2	3	4	5	6	7	8	9	10	11	12	13
CARDIFF	D.T. HERBERT 1970	1961	334	PCA	HT	HQ	EI SES	FS				
SWANSEA	D.T. HERBERT 1970	1961	221	PCA	HT	HQ	FS	E				
SUNDERLAND	B.T. ROBSON 1969	1961	263	PCA	SES	HQ	HC	FS	FS			
LEICESTER	W.K.D. DAVIS and G. LEWIS 1973	1966	139	PCA OBLIQUE ROTATION	SES	M	LC	HQ	MI LC	E	EP	UF
NOTTINGHAM	C.J.THOMAS 1968	1966	140	PCA	SES	HT and HQ	EP	SES	EP			
CARDIFF	D.J. EVANS 1973	1966	NOT STATED	PCA VARIMAX ROTATION	SES	LC/ HT	HQ	TI LC	FS			

Table 3.8 continued....

1	2	3	4	5	6	7	8	9	10	11	12	13
SWANSEA	D.J. EVANS 1973	1966	NOT STATED	PCA VARIMAX ROTATION	SES	LC/ T	HQ	TI LC	FS			
NEWPORT	D.J. EVANS 1973	1966	NOT STATED	PCA VARIMAX ROTATION	SES	T/ LC	LC	HQ	SES			
CARDIFF	J.W. RAINE 1976	1971	532	PCA OBLIQUE ROTATION	SES	HT	FS	HT	M	E	M	FS
GREATER NOTTINGHAM	C.R. VICTOR 1979	1971	983	PA OBLIQUE ROTATION	SES	HT	FS	LC	EP	E1 M	EP	HT1 FS
HULL	R.N. DAVIDSON 1975		1971	601	PCA	HT	HQ	LC	FS	EP	EP	

Key to Table 3.8:

PCA = Principal Components Analysis
PA = Principal Axis Factor Analysis
SES = Socio-Economic Status
HT = House Tenure
E = Ethnicity
FS = Family Status

HQ = Housing Quality
M = Mobility
LE = Life Cycle
EP = Economic Participation
UF = Urban Fringe
PS = Population Structure

should, therefore, be formulated in terms of the general dimensions of urban structure rather than the three constructs proposed by Shevky and Bell (1955).

Conclusion

This paper has demosntrated that the nature of urban social segregation in Greater Nottingham in 1971 could be described in terms of eight primary factors. By employing an oblique rotation procedure, it was demonstrated that these dimensions were inter-correlated. A higher order factor analysis illustrated that these dimensions could be summarised by three second order dimensions. Unfortunately, insufficient previous examples of the use of this technique existed for a comparison to be meaningful. Comparison of the primary order results with those produced by other workers proved useful. It was illustrated that two basic types of factor structure could be identified within the urban environment, these were the general and place specific elements. Socio-economic status, tenure, and family status—life-cycle stage were the three elements noted as general. The disparity between these elements and the three urban structure axes suggested by Shevky and Bell was noted and discussed. As a general conclusion it was considered that future theoretical developments in British social studies should be specific culturally and should be formulated in terms of the general elements of urban structure.

REFERENCES

Bell, W.; Economic, family and ethnic status; **American Sociological Review**; 20; (1955); 45-52.

Burgess, G.W.; The growth of the city: an introduction to a research project; **Publications of the American Sociological Society**; 18; (1924); 85-97.

Davidson, R.N.; **The Analysis of urban sub-areas**; S.S.R.C. Project Report; Department of Geography; Hull University; U.K.; (1975).

Davies, W.K.D.; Variance allocation and the dimensions of British towns; **Tijidschrift voor Economische en Sociale Geografie**; 66; 1975; 358-372.

Davies, W.K.D. and Lewis, G.; Urban dimensions in Leicester; In Clark, M. and Gleave, B., (eds.); **Social Patterns in Cities**; Special publication of the Institute of British Geographers; 5; (1973); 71-86.

E[illegible]ards, J., Leigh E. and Marshall, T.; Social patterns in Birmingham: 1966; **Occasional Paper 13**; Centre for Urban and Regional Studies; University of Birmingham; (1970).

E[illegible] D.J.; A comparative study of urban social structure in

Wales; In Clark, M. and Gleave B. (eds.); **Social Patterns in Cities**; Special publication no. 5; Institute of British Geographers; (1973); 87-102.

Giggs, J.A.; The distribution of schizophrenics in Nottingham; **Transactions Institute of British Geographers**; 59; (1973); 55-76.

Giggs, J.A. and Mather, P.M.; Factorial ecology and factor invariance: an investigation; **Economic Geography**; 51 (4); (1975); 366-382.

Harman, H.H.; **Modern factor analysis**; Chicago; University of Chicago Press; (1967).

Herbert, D.T., Principal components analysis and urban social structure: a study of Cardiff and Swansea; In Carter, H. and Davies, W.K.D. (eds.); **Urban Essays: Studies in the Geography of Wales**; London, Longman and Son; (1970); 79-100.

Hoyt, H., **The Structure and growth of residential neighbourhoods in American cities**; Federal Housing Administration, Washington D.C.; (1939).

Johnston, R.J.; Residential area characteristics: research methods for identifying urban sub-areas. In Herbert, D.T. and Johnston, R.J. (eds.); **Social areas in cities**; 1, 193-236; London; John Wiley and Son; (1976).

Kaiser, H.E.; The Varimax criterion for analytical rotation in factor analysis; **Psychometrika**; 23; (1958); 187-200.

Mather, P.M.; **Computational methods of multivariate analysis in physical geography**; London; John Wiley and Son; (1976).

Norman, P.; Third survey of London life: a new typology of London's districts. In Dokhan, M. and Rogan, R. (eds.); **Quantative ecological analysis in the social sciences**; Cambridge, Mass.; M.I.T. Press; (1969); 371-396.

Palm, R. and Caruso, D.; Labelling in factorial ecology; **Annals Association of American Geographers**; 62 (1972); 122-133.

Raine, J.W.; **Social interaction and urban neighbourhood: a case study of Cardiff**; unpublished Ph.D. thesis; University of Wales; (1976).

Rees, P.M.; Factorial ecology: an extended definition, survey and critique; **Economic Geography**; 47; (1971); 220-233.

Robson, B.T.; **Urban analysis**; Cambridge; Cambridge University Press; (1969).

Rummel, R.D.; **Factorial Analysis**; University of Chicago Press, Chicago; (1970).

Shevky, E. and Bell, W.; **Social area analysis: theory, illustr tive application and computational procedure**; Stanfor Calif.; Stanford University Press; (1955).

Smith, D.M.; **The geography of social well being**; New York Mcgraw Hill; (1973).

Thomas, C.J.; **Geographical aspects of the growth of the**

residential area of Greater Nottingham; unpublished Ph.D. thesis; University of Nottingham; (1968).

Victor, C.R.; **Social structure and social malaise: a geography of Greater Nottingham**; unpublished M.Phil. thesis; University of Nottingham; (1979).

———— **Social Area Change in Nottingham C.B.: 1961–1971**; Paper presented to the Institute of British Geographers Annual Conference; Lancaster; (1980).

FOUR

Erling Berge

Comparing the Factorial Ecology of Helsinki and Oslo

A recent review of substantive findings in Social Area Analysis and Factorial Ecology (Hamm, 1982) cites 78 studies involving a total of 66 cities. Of the 78 studies 41 are concerned exclusively with cities in the U.S.A. Only 9 studies report on the factorial ecology of cities in non-industrialised countries.

Scandinavian cities have got its share of studies. Most work has been done on Helsinki (Gronholm' 1961; Sweetser 1965a,c,1969,1973). Copenhagen (Pedersen, 1967) has been studied, and Janson (1971, 1976) by pooling data from several cities has studied the spatial structure of Swedish cities. Trondheim was studied by Dale (1981) and Oslo by Berge and Tember (1982). The present article reports the findings for Oslo and compares the factorial ecology of Oslo with that of Helsinki.

The Factorial Ecology of Oslo

Oslo is the capital city of Norway. By international standards it is a small city. In 1980 it had 452.023 inhabitants, in 1970 477.898 people lived there. The urban area, however, is larger.

----

* Eva Tamber assisted during the early phases of the study. The work of F.L. Sweetser has, as the observant reader will gather, been the primary inspiration. Data for the study were made available by the Norwegian Social Science Data Services. I am grateful for their contributions, but neither of them bear any responsibilities for the present use.

Adding the population of 9 surrounding municipalities to the population of Oslo shows that in 1980 a total of 698.283 people were living in an area which might be called "Greater Oslo" (see figure 4.1). In 1970 the area had only slightly fewer: 693.026 inh.

The present analysis is a study of "Greater Oslo" in 1970. But it does not include quite the 10 municipalities referred to above. The population and housing census of 1970 utilised 621 census tracts within these 10 municipalities. The tracts without population or only very sparsely populated at the outer edge of the area were removed so that 562 tracts were left covering a contiguous area mostly densely settled. Census tracts with less than 500 inhabitants were added to neighbouring tracts. This left 442 analytical units. Data on these 442 units were readily available from the Census tract data bank of the Norwegian Social Science Data Services. From these data 37 variables were computed. Their definitions are detailed in table 4.1. In order to compare the factorial ecology of Oslo with that of Helsinki an effort was made to replicate the variables used by Sweetser (1973, table 4.1) in his study of Helsinki. For 21 of the 33 variables used by Sweetser, the definitions are very similar. For another variable a reasonable approximation was found. The comparison of Oslo and Helsinki will be based on these.

Of the 37 variables those numbered 32-38 (var. no. 30 was discovered to contain random noise and excluded) were essentially uncorrelated with the others and according to established procedures (Sweetser, 1974) removed from the factor analysis. The remaining 30 variables (nos. 1-31) were then analysed by the principal factors method (PA2 option in the FACTOR program of SPSS: Nie, et al., 1975) and rotated to simple structure according to the varimax criterion. This gave the factor matrix with four factors reported in table 4.2.

Factor analysis does not produce any "best" solution automatically. Both the number of factors and the rotation to simple structure has to be decided upon by the researcher. Solutions with six and five factors were examined. Four factors seemed, however, to give the most parsimonious and meaningful description of the data. But evidence suggests that choice of factor model and criterion for rotation do not have much substantial impact (Berge, 1981, page 312-18, Hamm, 1979, p.31-36).

The four factors of table 4.2 account for 77.2 per cent of the variance of the observations. The strongest factor accounts for 31.3 per cent, the weakest only 6.9 per cent. Only one of the variables has communality less than .50. That is variable 7

FIG. 4.1 Small-scale map of selected units in Oslo and Akershus.

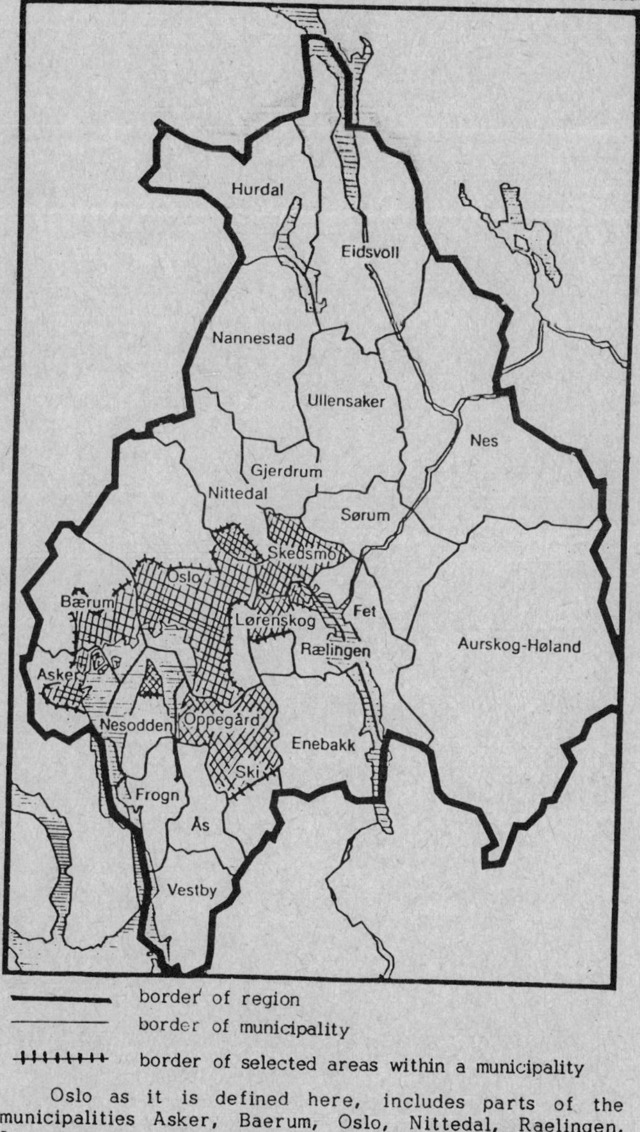

Oslo as it is defined here, includes parts of the municipalities Asker, Baerum, Oslo, Nittedal, Raelingen, Lorenskog, Ski and Nesodden, and all of Skedsmo and Oppegard.

"Regionplan for Oslo og Akershus". Regionplankontoret for Oslo og Akersus. Oslo 1976.

TABLE 4.1
Ecological Variables in Oslo and Helsinki
(transformations in parenthesis)

OSLO VARIABLES No.	Name	Definition	HELSINKI VARIABLES[b] No.	Name	Replic Status[a]
1.	Age 0-4	% of population age 0-4 years	2.	% 0-4 years	R-1
2.	Age 5-14	% of population age 5-14 years	3.	% 5-14 years	R-1
3.	Preadolescent ratio	% of population under 16 who are 0-4	4.	% u.15 who are 0-4	R-2
4.	Young middle age	% of population 20-69 who are20-39	5.	% who are 20-39	R-2
5.	Age 60 +	% of population who are 60 and over	6.	% who are 65 and over	R-2
6.	% male	% of population male	7.	% male	R-1
7.	Foreigners	% of population not born in Norway	8.	% speaking Swedish as principle language	A
8.	Married	% of population who are married	9.	% 15 years and older who are married	R-2
9.	Fertility ratio	Number of children 0-4 years per 1000 males age 20-49	10.	Number of children 0-4 per 1000 males 20-49	R-1
10.	Unmarried women	% unmarried women age 30-39	11.	% 15 years and older who are widowed or divorced	S
11.	One-person families	% of population over 16 years in one-person households	12.	% 15 years and older in one-person households	R-2

(a) R-1 = exact replication, R-2 = very close approximation, A = approximation, S = substitute, E = "extra" variable.
(b) Helsinki variables no. 1 (median age) and 15 (living space) could not be matched or ammproximated.

Table 4.1 continued....

OSLO VARIABLES No.	Name	Definition	HELSINKI VARIABLES[b] No.	Name	Replic Status[a]
12.	Size of household	Mean number of persons in families with two or more persons	13.	Mean number of persons in households with two or more persons	R-1
13.	Room crowding	% dwellings with room crowding less than 0.50 per room	14.	Number of persons per 100 rooms	S
14.	Home ownership	% of dwellings occupied by owners	16.	% of dwellings occupied by owners (incl. share-holders)	R-1
15.	Working women	% of women 16-69 years who are economically active (full-time)	17.	% of females age 15 and older who are economically active (excl. those unemployed and seeking work	R-2
16.	Manufacturing	% of economically active dependent on manufacturing industry ["naeringskode" 11-39, 51-52 (b)]	19.	Proportion of total population economically dependent on manufacturing industry (deciles)	R-2
17.	Service	% of economically active dependent on services ["naeringskoldes" 67-69, 81-93 (b)]	20.	Proportion of total population economically dependent on services, incl. transportation, commerce, etc. (declies)	R-2
18.	Employers	% of population bdY&g leaders of manufacturing a d organisations ["yrkeskode" 11 (c)]	21.	% of economically active who are employers of labor	S
19.	Blue collar	% of population who are manual workers ["yrkeskode" 50-59,	22.	% of economically active who are manual workers	R-1

(a) For definitions see above.
(b) Standard for naeringsgruppering i offentlig norsk statistikk. SSB Handbker nr. 9.
(c) Standard for yrkesgruppering i offentling norsk statistikk. Nordisk yrkesklassifisering. Arbeidsdirektoratet

Table 4.1 continued....

OSLO VARIABLES			HELSINKI VARIABLES[b]		
No.	Name	Definition	No.	Name	Replic Status[a]
20.	Primary education	% of population age 25-69 who have "folkeskole" education as their highest degree	23.	% of population age 15 and older who have passed middle school or student examinations	S
21.	Female higher education	Number of females with university education per 100 females with "gymnas"	24.	Number females who have passed student examination (for university admission) per 100 females who have passed middle school examination	R-1
22.	Apartment house size	% dwellings in building with 3 or more floors	28.	Number of dwellings per residential building	S
23.	Detached dwellings	% of dwellings in one or two dwelling buildings	29.	% of dwellings in one or two dwelling buildings	R-1
24.	Small dwellings	% of dwellings with one room (excluding kitchen)	30.	% of dwellings with one or two rooms (incl. kitchen)	R-1
25.	New Housing	% of dwellings built 1961-1970	31.	% of dwellings built 1951-1960	R-2
26.	Telephone	% of dwellings with telephone			E
27.	Dwellings without toilet	% of dwellings lacking toilet	33.	% of dwellings lacking flush toilet	R-1
28.	Sanitary standard	Dwellings lacking water, joint water conduit or sewerage (divided with total dwellings multiplied with 3. Square root)			E

(a) For definitions see above.
(b) He[illegible]i variables no. 25 (population density), no. 26 (Public buildings), no. 27. (Industrial buildings) and no. 32. ([illegible]ng defect heat) could not be matched or approximated.

Table 4.1 continued......

No.	OSLO VARIABLES Name	Definition	HELSINKI VARIABLES[b] No.	Name	Replic Status[a]
29.	Ec. active mothers	% women age 16-59 with children age 0-12 who are economically active (full time)			E
30.		random noise			
31.	Ec. active pensioner/ratio	Number of pensioners per 100 persons who are economically active			E
32.	(b) Agriculture	% of economically active dependent on agriculture. ("naeringskode" 01-02) (square root)	18.	Proportion of total population economically dependent on agriculture (deciles)	R-2
33.	Trade	% of economically active dependent on trade (naeringskode" 61-66)			E
34.	Total pop.	The total number of people (logarithm)			E
35.	Sex ratio	Number of women 16-39 years, per 100 men age 16-39 (square root)			E
36.	Dependency ratio	The population under 20 years plus the population above 60 years divided by the population between 20-59, all multiplied by hundred			E
37.	Non-family	% of one-person households with one person aged 30-66			E
38.	Work place	% of population who are occupied within commune of residence			E

(a) For definitions see above.
(b) Variables 32 to 38 were essentially uncorrelated with others in the matrix.

TABLE 4.2
Four Dimensions of Neighbourhood Differentiation, Oslo 1970
(Varimax Rotations of Principal Components solution)

Variables No.	Name	Replication Status[a)]	Factor Coefficients[b)] (Decimal Points Omitted) "Deprivation"	Socioeconomic Status	Familism	"Detached dwellings"	Communality
1.	Age 0-4	R-1	-07	-11	93	13	89
2.	Age 5-14	R-1	-67	11	57	27	86
3.	Preadolescent ratio	R-2	72	24	15	-19	64
4.	Young middle age	R-2	11	24	89	15	89
5.	Age 60+	R-2	50	-12	-68	-28	80
6.	% male	R-1	-56	-22	32	39	62
7.	Foreigners	A	21	65	01	12	48
8.	Married	R-2	-69	-32	06	06	58
9.	Fertility ratio	R-1	04	-15	78	07	63
10.	Unmarried women	S	73	04	-37	-37	81
11.	One-person families	R-2	86	-04	-27	-28	89
12.	Size of household	R-1	-59	28	56	37	88
13.	Room crowding	A	-08	70	-34	18	64
14.	Home ownership	R-1	-43	32	15	77	89
15.	Working women	R-2	67	23	-12	-39	66
16.	Manufacturing	R-2	-12	-81	04	-06	67
17.	Services	R-2	14	77	17	-21	67

Table 4.2 continued....

18. Employees	S	-35	74	-13	26	76
19. Blue collar	R-1	16	-95	-06	-03	93
20. Primary educ.	A	34	-88	-20	-11	95
21. Female higher educ.	R-1	-09	73	17	17	59
22. Apartment house size	S	46	-26	-21	-74	87
23. Detached dwellings	R-1	-40	24	12	81	89
24. Small dwellings	R-1	80	-19	-11	-17	72
25. New housing	R-2	-25	14	78	-05	69
26. Telephone	E	-59	67	-32	-08	91
27. Dwellings without toilet	R-1	65	-35	-02	29	63
28. Sanitary standard	E	-04	01	12	70	50
29. Ec. active mothers	E	-26	-12	72	01	60
31. Ec. active/pensioner ratio	E	41	-02	-54	-23	50
FACTOR VARIANCE		11.67	5.89	3.50	1.85	
% OF TOTAL VARIANCE		38.3	19.9	12.1	6.9	77.±

a) R-1 = exact replication, R-2 = very close replication, A = approximation, S = substitute, E = "extra" variable.
b) Coefficient of ±, 45 or more in parentheses ______

(Foreigners) with 0.48. Total of 26 variables have communalities above 0.60. Table 4.2 presents the factor loadings of the variables on the four factors along with communalities and factor variances. Factor loadings with absolute value of 0.45 or more have been put in parenthesis. They are usually taken to be large enough to be of interest in judging the nature of the social relations the factor is a measure of. In table 4.3 the variables have been arranged according to the size of the loading for each of the four factors.

Interpretation of the analysis

Factorial ecology (see f.e. Timms, 1971 or Sweetser, 1982) has established beyond doubt the general validity of the three dimensions hypothesised by social area analysis (Shevky and Williams, 1949; Shevky and Bell; 1955). These dimensions are now usually labelled socio-economic status, ethnicity and familism. And they are taken to be basic dimensions of the social structure.

The present writer has argued (Berge, 1981, 1982; Berge and Tamber, 1982) that a factor identifies the relative strength of one position in a dichotomized dimension of the social structure.

Basically there are two types of social structural dimensions. One type will when dichotomized, have a top-position and a bottom position. The other type will have a member position and a non-member position. The first type might be called inequality dimensions, the second one equality dimensions.

The first factor is by far the strongest. It accounts for 38.3 per cent of the variance of the data. This factor is dominated by variables like "one person families" and "small dwellings". The old people and single women might suggest a life cycle factor, but the small dwellings and the relative lack of telephones as well as the relative scarcity of men suggests the importance of material conditions of living and the relative deprivation of the population described by these variables.

The distribution of material conditions of living has been shown to be more sqewed in cities than in the country as a whole (Aase and Dale, 1978). It, therefore, seems right to stress the aspects of living conditions suggested by the factor.

The dimension of the social structure identified by this factor is an inequality dimension identifying the relative strength of a "have not" position, and, by logical implication, the "have"

TABLE 4.3
Four Ecological Factors, Oslo 1970
(Decimal points omitted)

Variable No.	Name[a]	Factor +	Coefficient -
	FACTOR I "DEPRIVATION"		
11.	One-person families	86	
24.	Small dwellings	80	
10.	Unmarried women	73	
3.	Preadoelscent ratio	72	
8.	Maried		-69
15.	Working women	67	
2.	Age 5-14		-67
27.	Dwellings without toilet	65	
12.	Size of household		-59
26.	Telephone		-59
6.	% male		-56
5.	Age 60+	50	
22.	Apartment house size	46	
	FACTOR III "FAMILISM"		
1.	Age 0-4	93	
4.	Young middle age	89	
9.	Fertility ratio	78	
25.	New housing	78	
29.	Ec. active mothers	72	
5.	Age 60+		-68
2.	Age 5-14	57	
12.	Size of household	56	
31.	Ec. active pensioner ratio		-54

Variable No.	Name	Factor +	Coefficient -
	FACTOR II "SOSIOECONOMIC STATUS"		
19.	Blue collar		-95
20.	Primary educ.		-88
16.	Manufacturing		-81
17.	Service	77	
18.	Employers	74	
21.	Female higher educ.	73	
13.	Room crowding	70	
26.	Telephone	67	
7.	Foreigners	65	
	FACTOR IV "DETACHED DWELLINGS"		
23.	Datached dwellings	81	
14.	Home ownership	77	
22.	Apartment house size		-74
28.	Sanitary standard	70	

(a)For further definitions see table 1.

position. This suggests the name "Deprivation" for this factor.

The second strongest factor is dominated by variables like "proportion of the economically active population in service occupations", "self-employed people" and "women with high education" with positive loadings and "blue collar workers" and "proportion with primary education" with negative loadings. The aggregates described by these variables seem to be dominated by inequalities in educational and occupational resources. The factor analysis identifies the relative strength of a top position in a dichotomized structure where presence of socio-economic status resources suggest the name:"Socio-economic Status" for the factor. The one variable which might indicate otherwise is "porportion not borne in Norway" since immigrants usually will be low status workers. But before 1970 most immigrants to Norway were high status workers from the other Scandinavian countries and the U.S.A.

The third factor is clearly a life cycle factor. It is dominated by variables like "proportion of population aged 0-4", "proportion of population aged 20-69 who are of age 20-39", "fertility ratio" and "new dwellings".

The aggregates described by these variables are related to the social structure in a way different from the former two factors. While the former two factors described hierachical structures, the present factor seems to indicate a structure of the membership/non-membership type. The variables loading high on the factor seem to suggest that the position identified, is the "standard" complete nuclear family. Hence the label "familism" for this factor.

The fourth factor is very weakly defined, accounting for only 6.9 per cent of the variance. Only four variables loads on the factor and all of them concern housing. The variables seem to have something to do with low standard rural dwellings, and might suggest the relative strength of the "native" population of a census tract or the degree to which the area was used for holiday cabins before city growth made the area a part of the city. If the first possibility is the case, the factor is of the same type as the Familism factor. It reports the relative strength of a member position. The second possibility would make the factor a candidate for exclusion from the analysis. It would then describe the environment of the social structure proper (Berge, 1982). In order to resolve this question one would have to include additional data in a new factor analysis. Lacking these data the case must rest unresolved.

The factor scores on the three major dimensions identified, measure the strength of a position in the social structure relative to the average strength of the position within the area studied. One of the more interesting results of a factor analysis like the present one, is thus the picture it offers of the spatial distribution of the social structure.

The city of Oslo is divided into 60 zones. Taking the average of the factor scores on the analytical units within each zone and dividing these zone scores into 6 size classes make it possible to show the spatial distribution of structural positions on a map of Oslo by giving each size class a shade of gray.

In figures 4.2 - 4.4 this is done for the three major dimensions identified. Dark shading indicates high factor score values. Deprivation is most clearly present in the east part of the central city. Socio-economic status is relatively highest in the western parts of the city and familism strongest on the north-eastern part and the municipalities outside of Oslo (not shown on map).

If spatial clustering is a general tendency for actors in the same position in the social structure, factor scores on census tracts within zones ought to be more like each other than scores on tracts in other zones. There ought to be less variation in factor score values within zones than between zones. A test of variances confirms the general tendency for spatial clustering. (The F-statistic for the Deprivation factor is 8.6, for the Socio-economic status factor it is 14.5 and for the Familism factor it is 4.2 with 62 and 379 degrees of freedom.)

The general validity of the three societal dimensions— Socio-economic status, Familism/Urbanism and Ethnicity/Immigrant status—proposed by Shevky and his colleagues is well documented (Hamm, 1982). The discovery of a Socio-economic status factor and a Familism factor in Oslo is, therefore, no surprise. Neither is the absence of an ethnicity/immigrant status factor surprising (Sweetser, 1969). The interesting fact here is the appearance of a "Deprivation" factor as the major factor of differentiation in the Oslo metropolitan region. A comparison of the factor pattern in Oslo with the pattern in Helsinki may help clarify the meaning of the Deprivation factor.

Comparing Oslo and Helsinki

Of the 21 variables with similar definitions in this study of Oslo and Sweetser's study of Helsinki, one dropped out of the analysis because of low correlations with other variables. Of

FIG. 4.2 Geographical location of various neighbourhood types. (mean of factor scores).

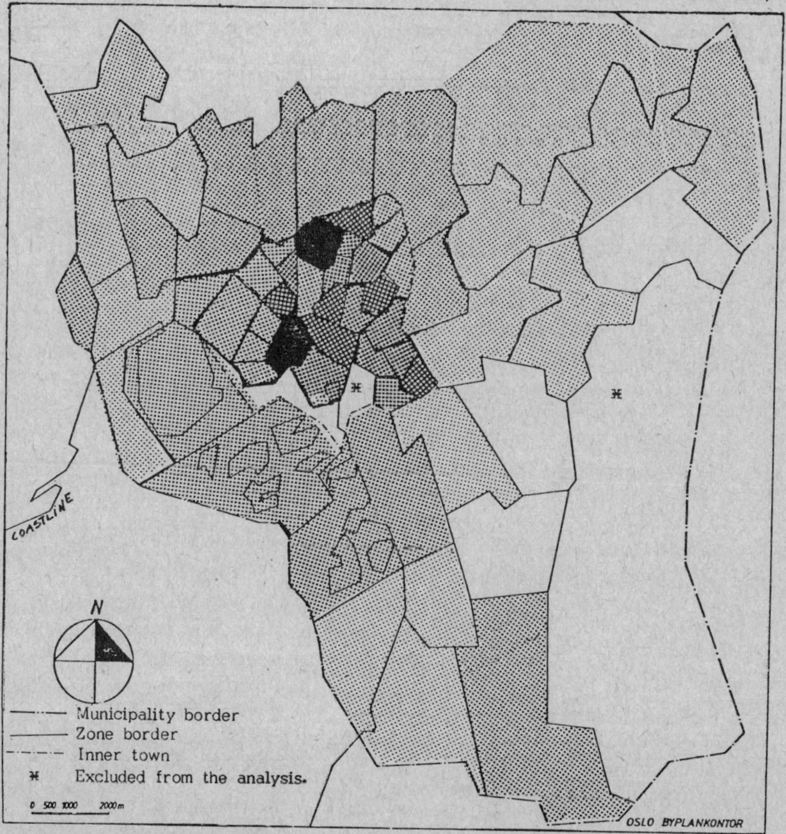

FACTOR I "Deprivation". Oslo.

Dark shadings = high values.

FIG. 4.3 Geographical location of various neighbourhood types (mean of factor scores).

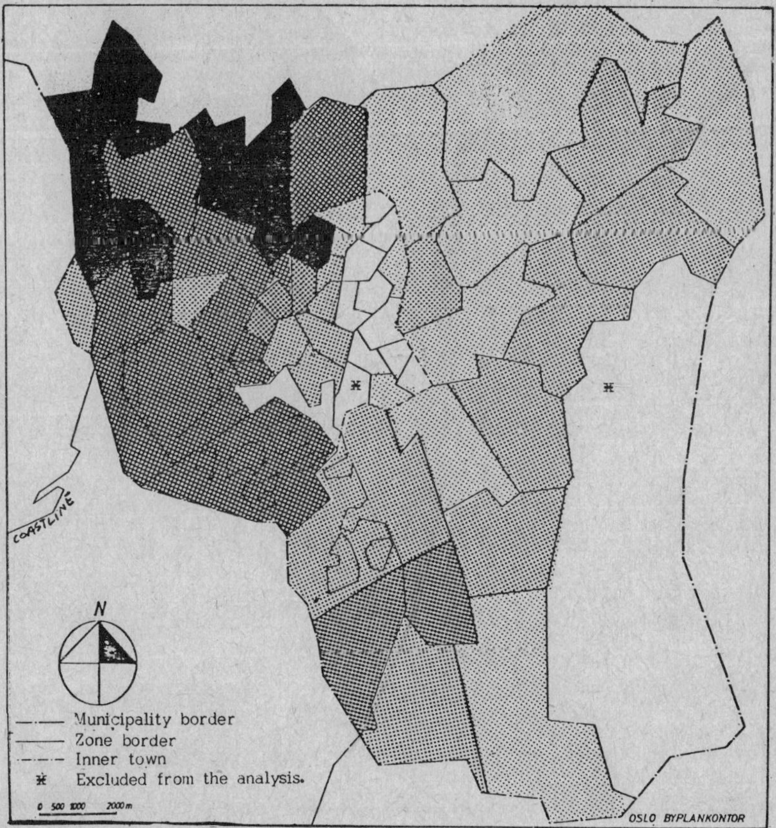

FACTOR II "Socio-Economic Status (SES)". Oslo.

Dark shadings = high values

FIG. 4.4. Geographical location of various neighbourhood types. (mean of factor scores).

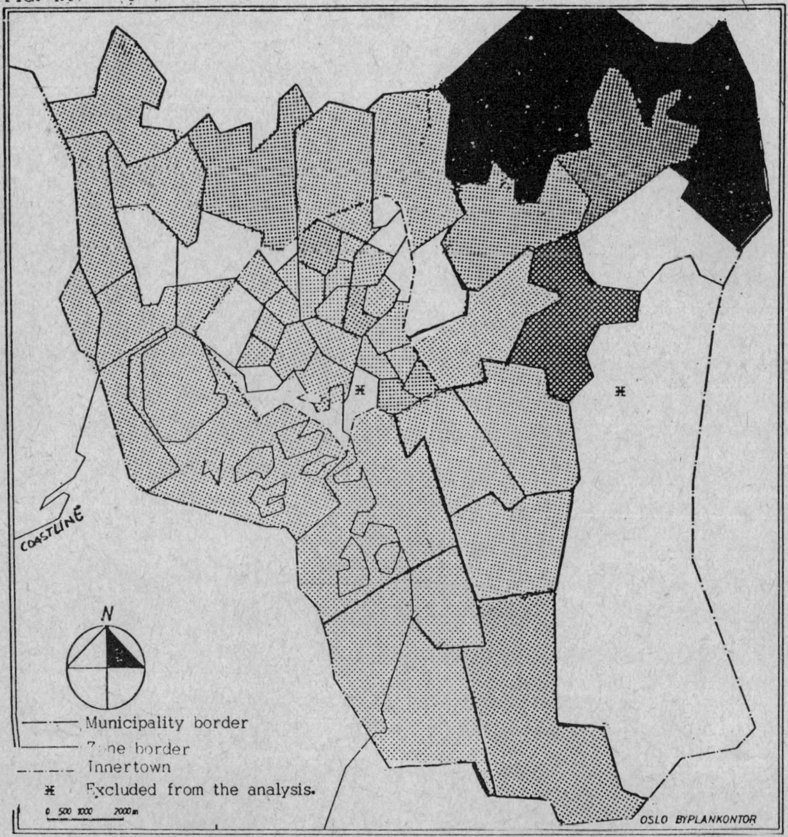

Factor III "Familism". Oslo.

Dark shadings = high values.

the approximations only one per cent of population not born in Norway (approximation of per cent Swedish speaking Finns), could be included. The others were defined reciprocally in relation to the variable they were approximations of. Table 4.4 is taken from table 4.14 in Sweetser, 1973 and gives for the comparable variables the factor loadings on the six factors identified by Sweetser as describing the social structure of Helsinki. From tables 4.2 and 4.4 one may compute the coefficient of factor congruence (Harmon, 1967, pp. 269-271) between factors identified in Oslo and Helsinki. The results are reported in table 4.5. In table 4.6 the factors from the two cities are compared variable by variable.

The Socio-economic status factor is clearly the same in both cities. From table 4.6 it is seen that the largest differences are found for the variables "detached dwellings" and "foreigners". These differences are easily explained since Oslo had "Detached dwellings" as a separate factor and Helsinki had "Swedish language".

The Familism and Deprivation factors of Oslo are each related to two of the Helsinki factors.

Coefficients of Congruence

OSLO	Familism		Deprivation	
	.83	-.77	.64	-.61
HELSINKI	Young Familism	Postgeniture		Familism/ Ruralism

The Familism factor in Oslo resembles both Young Familism and Postgeniture in Helsinki. Postgeniture also resembles the Deprivation factor in Oslo and so does the Familism-Ruralism factor. The picture is a bit complicated, but it is not random. The factors are closely related, but even so the conclusion must be that the factorial ecology of Oslo, 1970 is different from Helsinki's in 1960.

The comparison has suggested that the Deprivation factor of Oslo is some kind of family cycle factor. This is a contradiction of the interpretation of this factor as an inequality dimension and deserves some further investigation.

TABLE 4.4

Six Dimensions of Neighbourhood differentiation, Helsinki Region 1960. Factor Coefficients of 21 Variables replicated in Oslo[a] (Varimax rotations of Principle Components Solution.)

No.	Name	Socio-Economic Status	Postgeniture	Familism Ruralism	Young Familism	Residentialism	Swedisn Language
		Factor Coefficients (Decimal Points Omitted)					
1.	Age 0-4	-27	-50	17	75	23	-02
2.	Age 5-14	-12	-51	68	24	01	-20
3.	Preadolescent ratio	-26	-28	-27	72	29	11
4.	Middle age ratio	-07	-57	-08	67	15	-12
5.	Age 65+	14	87	16	-20	-08	15
6.	Prop. male	-43	-72	28	06	04	03
7.	Swedish speaking Finns	29	20	07	02	01	72
8.	Married	-29	-74	11	34	34	-02
9.	Fertility ratio	-21	-21	40	80	16	01
11.	One-person families	19	57	-42	-38	-44	-08
12.	Size of household	02	-49	73	31	03	-04
14.	Home ownership	10	-33	34	02	73	-15
15.	Working women	22	11	-73	-05	-29	-34
16.	Manufacturing	-76	-44	-01	15	10	-15
17.	Services	76	06	-46	-05	-09	02
19.	Blue collar	-95	-14	-01	04	-08	00
21.	Female higher educ.	64	05	09	-03	-06	-03
23.	Detached dwellings	-47	-33	59	09	39	30
24.	Small dwellings	-59	52	-26	10	-28	-32
25.	New housing	17	-61	-09	39	41	-29
27.	Dwellings without toilet	-70	-21	46	19	20	25
	Factor variance (33 variables)	6.88	6.33	5.78	3.56	2.63	1.98
	% of total variance (33 variables)	20.9	19.2	17.5	10.8	8.0	6.0

[a] Sweetser, F. L. (1973) page 45. See also table 1 above.

TABLE 4.5
Coefficients of Congruence[a] between Factors Derived from Oslo Analysis, 1970 and Factors Derived from Metropolitan analysis, Helsinki, 1960.
(21 Matched Variables)

Helsinki factors	Oslo factors: Deprivation	Socio-Economic Status	Familism
Socio-Economic Status	-.02	.83	-.11
Postgeniture	.64	.12	-.77
Familism-Ruralism	-.61	-.04	.37
Young Familism	-.11	-.13	.83
Residentialism	-.48	.01	.48
Swedish Language	.04	.17	-.21

[a]Cf. Harmon, H. H. (1967) page 269-271.

To aid the reinterpretation it will be compared to the reflected Familism/Urbanism factor of the "Urban Residential areas in Australia" (Sweetser, 1982, Appendix table 1).

Variable by variable for all variables correlation with an absolute value of .50 or more with either of the two factors, it came out like shown in table 4.7.

TABLE 4.6
Comparison of factor loadings from similar factors in Helsinki and Oslo Variable

No.	Name	Oslo	Helsinki status
SOCIO-ECONOMIC STATUS			
17. Dependent on services	.77	.76	R-2
21. Female high education	.73	.64	R-1
07. Foreigners	.65	.29	A
23. Detached dwellings	.24	-.47	R-1
24. Small dwellings	-.19	-.59	R-1
27. HU's without toilet	-.35	-.70	R-1
16. Dependent on manufacturing	-.81	-.76	R-2
19. Blue collar	-.95	-.95	R-1
COEFFICIENT OF CONGRUENCE		.83	

FAMILISM		Post-gen.	Young Fam.	
1. Age 0-4	.93	-.50	.75	R-1
4. Young middle age	.89	-.57	.67	R-2
9. Fertility ratio	.78	-.21	.80	R-1
25. New housing	.78	-.61	.39	R-2
2. Age 5-14	.57	-.51	.24	R-1
12. Size of HH	.56	-.49	.31	R-1
6. % male	.32	-.72	.06	R-1
3. Preadolescent ratio	.15	-.28	.72	R-2
8. % Married	.06	-.74	.34	R-2
24. Small dwellings	-.11	.52	.10	R-1
11. One person families	-.27	.57	-.38	R-2
5. Age 60+	-.68	.87	-.20	R-2
COEFFICIENT OF CONGRUENCE		-.77	.83	

DEPRIVATION		Post-gen.	Fam/Rur.	
11. One person families	.86	.57	-.42	R-2
24. Small dwellings	.80	.52	-.26	R-1
3. Preadolescent ratio	.72	-.28	-.27	R-2
15. Working women	.67	.11	-.73	R-2
27. HU's without toilet	.65	-.21	.46	R-1
5. Age 60+	.50	.87	.16	R-2
17. Dependent on services	.14	.06	-.46	R-2
4. Young middle age	.11	-.57	-.08	R-2
1. Age 0-4	-.07	-.50	.17	R-1
25. New housing	-.25	-.61	-.09	R-2
23. Detached dwellings	-.40	-.33	.59	R-1
6. % male	-.56	-.72	.28	R-1
12. Size of HH	-.59	-.49	.73	R-1
2. Age 5-14	-.67	-.51	.68	R-1
8. % married	-.69	-.74	.11	R-2
COEFFICIENT OF CONGRUENCE		.64	-.61	

Comment:

If the definition of the Australian variables are close to the Oslo variables or the definition rather obvious from the short description used here, no more is noted. The important differences are as follows:

(1) The Oslo variables are married and unmarried women. If the complement had been used only the sign of the coefficient would change. Hence they are included in the comparison with signs changed.

(2) The Oslo variables were defined as Age 5-14 and Age 60+.

TABLE 4.7

Coefficient of correlation between variables and factors: the Deprivation factor of Oslo and the Familism/Urbanism factor (reflected) of the "Urban Residential Areas of Australia"

	Familism/ Urbanism -Australia	Deprivation Oslo
One-person HH	.92	.86
HH size	-.89	-.59
Married women 1)	-.88	-.73
Detached dwellings	-.88	-.40
Age 5-10	-.85	
Separated or divorced	.85	
HH with unrelated members	.85	
HH with no automobile	.84	
Small dwellings	.83	.80
Married men	-.77	
HH population in flats 4)	.76	
Age 11-15 2)	-.73	-.67
Preadolesent ratio 5)		.72
Age 65+ 2)	.68	.50
Never married 1)	.66	.69
Fertility ratio 3)	-.66	
Dwellings without toilet		.65
Women in labour force	.65	.67
Owner occupied dwellings	-.64	-.43
Private kitchen & bath	-.63	
Telephone		-.59
% male		-.56
Persons per 100 rooms	-.56	
HH with TV	-.50	
Large apartment bldg. 4)	.50	.46

(3) The Fertility ratio in Oslo was defined as children 0-4 per 100 men (like in Helsinki) and was not found to be comparable.

(4) The definitions of the Australian variables are as follows: "% of HH population in 2 flat bldg." and "% of flat population in bldg. with 9 flats". The Oslo variable apartment house size is taken to indicate roughly the same as the Australian Large Apartment Buildings.

(5) The Australian variable "primary-secondary ratio" was defined as age 5-10 as proportion of age 5-15 and not

judged comparable.

Of the 25 variables included in table 4.7 only 11 are in any way comparable. For these 11 the coefficient of factor congruence (Harman 1967, pp. 270) can be computed to 0.98. This only confirms the striking similarities which can be observed in the table.

The factor labelled "Deprivation" in Oslo is closely related to the factors usually found and identified as Familism/Urbanism factors. The Oslo factor should perhaps have been labelled Urbanism if tradition were to be followed. But looking again at table 4.7, it is seen that one third of the variables are direct indicators of a distribution of material standards of living and the others are more or less all indicators of the distribution of (dis)advantaged groups. Rather than weaken the interpretation of the factor as indicating a state of relative deprivation the comparison seems to support it. This indicates that the traditional interpretation of the factor may have focused on the wrong aspects when familism aspects were emphasised. It is suggested here that a better label would be Affluence/Deprivation instead of Familism/Urbanism.

Returning then to the comparison of Oslo and Helsinki, one may ask why no clear Familism/Urbanism or Affluence/Deprivation factor emerges? One reason may be related to the problem raised by Sweetser (1969, pp. 45) in a comparison of Boston and Helsinki. If care is not shown in choosing the boundaries of the study area, either inner city differentiation will dominate the date (if the study area is too small) or urban-rural differentiation will dominate the data (if the study are is too wide). The Familism/Ruralism factor of the Helsinki metropolitan area suggest that the latter has happened. Presumably removing the rural fringe from the metropolitan area should have resultated in two factors, one Familism factor and one Deprivation/Affluence (or Urbanism/Familism).

Conclusions

The factorial ecology of metropolitan Oslo in 1970 revealed clearly three dimensions. Two were found to be inequality dimensions and labelled Socio-economic status and Deprivation. A third was found to be an equality dimension and labelled Familism.

A comparison of the factors in Oslo with factors found in metropolitan Helsinki in 1960 (Sweetser, 1965,a.b.,1969) showed the Socio-economic status factors to be the same. But it turned up a problem in the interpretation of the Deprivation factor. Comparing the Deprivation factor of Oslo with the Familism/Urbanism factor of the Australian urban residental areas (Sweetser, 1982) suggested that the traditional interpretation of the Familism/Urbanism factor has overlooked its relation to distributions of material standards of living. It is suggested that a better label for this factor would be Affluence/Deprivation.

The reason for the absence of this factor in the Helsinki study is suggested to be a relative domination of rural/urban differentiation in the data used due to too inclusive definition of the study area.

REFERENCES

Aase, Asbjorn and Britt Dale; **Levekar i storby.** (The standard of living in cities); Norges Offentlige Utredninger; Oslo; Universitetsforlaget; 1978.

Arbeidsdirktoratet; "Nordisk Yrkesklassifisering. Standard for yrkesklassifisering i offerntlig norsk statistikk. "(Nordic standard occupational classification.) Oslo; Arbeidsdirektoratet; 1965.

Berge, Erling; "The Social Ecology of Human Fertility in Norway 1970". Ann Arbor. University Microfilms International; 1981.

——— "On the correlation of Factors in Factorial Ecology." Paper presented to the Committee on Social Ecology at the X World Congress of Sociology in Mexico City; Mexico; August, 1982.

Berge, Erling and Eva Tamber; "Oslo-regionens sosialeokologi." (The Social Ecology of the Oslo Region.) Oslo; Institute of Applied Social Research; 1982.

Dale, Beitt; "Faktor kologiske analyser - et hjelpemiddel i planlegginsorientert byforskning?" (Factorial Ecology - an aid in planning oriented urban research?) Arbeidsnotat nr 1, RFSP's Byforskningsporgram. **Naermilj prosjekted.** Senter for samfunnsforskning; Trondheim; Universitetet i Trondheim; 1981.

Gronholm, Leo; "The Ecology of Social Disorganization in Helsinki." **Acta Sociologica**; Vol. 5. No.1; page 31-40.

Hamm, Bernd; **Indikatoren der Stadtentwicklung.** Trier Beitrage zur Stadt & Regionplanung; Trier; Universitat Trier; 1979.

——— "Social Area Analysis and Factorial Ecology. A Review of Substantive Findings." in G.A.Theodorson : **Urban Patterns: Studies in Human Ecology.** University Park; The Pennsylvania State University Press; 1982.

Harman, Harry H; "Modern Factor Analysis" (2nd ed.) Chicago;

University of Chicago Press; 1967.

Janson, Carl Gunnar; "A preliminary Report on Swedish Urban Spatial Structure." **Economic Geography.** Vol. 47; 1971; page 249-257.

———"Swedish Urban Spatial Structure: 1965." Bulletin From The Swedish Institute for Building Research, 5; (1976); 23 pages.

Nie, Norman H. C. Hadlai Hull. Jean G. Jenkins. Karin Steinbrenner and Dale H. Bent; "SPSS. Statistical Package for the Social Sciences" (2nd ed.) New York McGraw-Hill; 1975.

Pedersen, P.O.; "Urban Population Structure in Copenhagen." Proceedings of the First Scandinavian-Polish Regional Science Seminar; Warsaw; Polish Scientific Publishers; 1967.

Regionplankontoret for Oslo og Akershus; "Regionplan for Oslo og Akershus" (Regional Plan for Oslo and Akershus) Oslo; Regionplankontoret for Oslo og Akershus; 1976.

Shevky, Eshref and Marilyn Williams; "The Social Areas of Los Angeles." Berkeley; University of California Press; 1949.

Shevky, Eshref and Wendell Bell; "Social Area Analysis." Stanford; Stanford University Press; 1955.

Statistisk Sentralbyra; "Standard for nringsgruppering." (Standard industiral classification) Handbker nr 9 Oslo; Central Bureau of Statistics; 1972.

Sweetser, Frank L.; "Factor Structure as Ecological Structure in Helsinki and Boston". **Acta Sociologica**; Vol. 8; 1965a; page 205-225.

———"Factorial Ecology Helsinki 1960; **Demography**; Vol.2, 1965b; page 372-385.

———"Ecological Factors in Metropolitan Zones and Sectors." in Dogan & Rokkan (eds.) : **Quantitative Ecological Analysis in the Social Sciences**: Cambridge; M.I.T.; Press; 1969

———"Metropolitan and Regional Social Ecology of Helsinki."**Commentationes Scientiarum Socialium**; 5,1973. Helsinki; Societas Scientiarum Fennica; 1973.

———"The Uses of Factorial Ecology in Classification." in Archer (ed.) : **Current Research in Sociology**; The Hague; Mouton; 1974.

———**Urban Residential Areas in Australia.** Canberra; The Australian National University; 1982.

Timms, Duncan W.; "The Urban Mosaic"; Cambridge; Cambridge University Press; (1971).

FIVE

Christina R. Victor

A Multivariate Analysis of Social Area Change in Nottingham C.B. 1961-1971

Spatial studies of the internal social formation of British cities have, like studies of Western cities generally, been largely built around the theoretical suggestions of social area analysis (Shevky and Bell, 1955) and the empirical findings of previous factorial ecologies (Davis and Lewis, 1974). In the North American context most factorial ecologies have confirmed the presence of Shevky and Bell's three basic dimensions of social structure (Sweetser, 1965), although some modifications have been identified (Johnston, 1973). Evidence to support these structural generalisations has been more ambiguous in Britain.

Whilst social area analyses and factorial ecologies have provided generalisations concerning the nature of residential segregation in cities, they do not directly relate to spatial patterns. Models of the latter are associated with the writings of Burgess (1924) and Hoyt (1939). However, several researchers have linked these spatial models with the findings of factorial ecologies, indicating that socio-economic status has a generally sectoral distribution, family status a zonal distribution and a cluster type of pattern for ethnic minority groups. (Timms, 1970 b). Again evidence in support of these particular generalisations is more ambiguous in the British urban context. (Robson, 1975).

Considerable energy has been expended by research workers in identifying social patterns in cities and their relationship to the Burgess and Hoyt models. As Johnston (1973) notes, few workers have investigated these models as generalisations of social change. As originally constituted, these spatial models were formulated as dynamic entities and not as generalisations relating to a single point in time. Briefly summarised, it was proposed that ethnic areas would expand outwards in a zonal fashion but within distinctive sectors. No decrease in segregation levels would accompany this expansion (Morril, 1965). An outward zonal movement of status groups was proposed by Burgess (1924). Hoyt (1939) considered this movement would also occur within sectors. According to the family status models areas "age" over time as their residents move through the stages of the family life cycle. Expansion of family status is predominantly zonal in orientation with new families tending to be established upon the periphery of the city.

Whilst these models are broad generalisations they provide little specific information concerning rates of change and their intra-urban variations. Individual studies of aspects of change in specific cities have been reported for both North America (Curson, 1967; Duncan and Duncan, 1957) and Britain (Barke, 1977; Davidson, 1977). As yet, however, little work has been published which has investigated the nature and pattern of social change in the urban social environment at a city wide scale of analysis. Recently Murdie (1969) has reported a study of social change in Toronto and Brown and Horton (1970) reported a similar exercise for Chicago. In general their findings, within the constraints of the data, supported the theories of the above models, whilst Johnston's (1973) study of Melbourne stressed the importance of immigration in influencing social change.

The present paper reports an attempt to use the technique of factor analysis as a means by which to identify change in the structure and pattern of social segregation in Nottingham C.B. between 1961 and 1971. The analyses are discussed in two main sections. The first replicates Brown and Horton's methodology whilst the second replicates an alternative analytical procedure suggested by Johnston (1973). The results are then compared with the postulates of the social change models and their applicability as predictors of social change discussed.

Problems involved in British temporal studies of urban social structure

That time dynamic studies of social area change have not been forthcoming in Britain may be largely attributed to three major factors. Firstly, examination of the individual census forms reveals that the same questions are not always included on successive questionnaires. For example, information relating to those persons who had received a formal education beyond the age of 16 was collected for the 1961 enumeration but not since. A second difficulty is posed by the often subtle changes in the definition of similar questions at different censuses. This often makes comparison of the derived variables hazardous. For the urban social geographer a third major difficulty has been the redefinition of the spatial framework with each successive census. As a direct result British workers have rarely been able to define a stable area framework within which to examine changes in socio-spatial differentiation over time.

Data

For this investigation the returns of the 1961 and 1971 censuses provided the data base with the pre-1974 County Borough of Nottingham defining the study area. Eighteen directly comparable variables were selected for inclusion in this analysis (Table 5.1). Four of these indices represented measures of the demographic structure; two were measures of ethnic status, a further two identified household structure, with a further four each measuring housing tenure and housing quality. Each of these indices were then recalculated for the 140 E.D's defined in the study area at the time of the 1966 10 per cent sample census. This spatial framework proved to be a direct aggregation of the 483 E.D's defined in 1961 and the 596 districts defined for the 1971 census.

TABLE 5.1

Variables used in the Comparative Factorial Ecology of Social Structure - 1961-1971

1. % Households owner occupiers.
2. % Households Local Authority renters.
3. % Households private rented (unfurnished).
4. % Households private rented (furnished).
5. % Households share/lack hot water tap.
6. % Households share/lack fixed bath/shower.
7. % Households share/lack W.C.
8. % Dwellings 1-3 rooms.
9. % Households sharing a dwelling.
10. % Households - single person.
11. % Households - 6+ persons.
12. % Households overcrowded (1.5+ persons per room).
13. % Population aged 0-14.
14. % Population aged 65+.
15. % Population aged 20-44.
16. % Population aged 45-60.
17. % Population foreign-born white (i.e., Ireland and Europe).
18. % Population foreign-born coloured (i.e., New Commonwealth

The dimensions of social change: a longitudinal approach

Brown and Horton (1970) suggested that a longitudinal analysis was the most appropriate means of identifying change over time in the social structure of cities. This methodology involves calculating the exact change in each variable between 1961 and 1971. This information is then reviewed by factor analysis to identify the dimensions of social change. Change coefficients were computed for each variable in each district using the formula:

$$cij = \frac{Pij\ (1961)}{Pij\ (1971)} \times \frac{100}{1}$$

where Pij (1961) = value for district i on variable j in 1961.

Pij (1971) = value for district i on variable j in 1971.

and Cij = change coefficient for district i on variabile j.

The resultant 18 x 140 data matrix was then analysed by a principal axes factoring routine. Eight factors with eigenvalues over one, representing 68.2 per cent of the initial variation, were extracted. Simple structure was obtained by rotating the factor axes according to the direct oblimin (oblique) rotation procedure, with delta set at zero. This yielded the pattern of loadings listed in Table 5.2. (Loadings of $\pm$ 0.3 or greater are underlined to aid interpretation in this and subsequent tables.) From these loadings eight factors can be interpreted as follows:

(1) A tenure-housing quality factor links low quality housing with the private rented (unfurnished) sector of the housing market. In contrast local authority housing is negatively associated with low housing quality.

(2) A life cycle status factor with positive loadings on he young adults (20-44 years) and negative loadings on the middle age (45-60) and private rented tenure variables.

(3) A tenure factor which contrasts home ownership with local authority rented tenure.

(4) A shared dwellings-amenities measure which contrasts shared dwellings and statutory overcrowding with the exclusive use of a hot water system.

(5) An elderly population factor which illustrates a positive loading with the measure of persons aged 65 and over. In contrast, large households (those containing

six or more persons) have a negative loading.

(6) A population growth factor which links increases in the total population with the growth of single person households.

(7) A deprivation factor links the coloured immigrant population with the young dependent population and the lack of housing amenities.

(8) A private renting factor which positively associates the private (furnished) rented sector of the housing market with the multi-occupation of dwellings and young children.

The classical models of urban social structure, outlined previously, suggested that status, family status and ethnic origions were the three basic aspects of urban social structure which would summarise social change. Clearly the eight factors identified in this example do not fit neatly into these three categories. Of the identified dimensions one relates to the concept of deprivation (factor 7), one to population growth (factor 6); two to life cycle status (factors 2 and 5) and four to housing (factors 1,3,4 and 8). Housing would appear, therefore, to have been the most dynamic element of the social environment of Nottingham during the specified time period. That this investigation did not validate the primacy of the dimensions of change postulated by the residential models may well reflect the cultural differences experienced between North America (the social context in which the models were formulated) and Britain.

It is, however, important to stress that this specific analytical procedure illustrates a number of limitations which lessen its utility as a tool for identifying social change in cities. Firstly this is a deterministic mode of analysis. A situation of change is imposed upon the factor solution by the nature of the primary variables as measures of social change. Such a methodology precludes the identification of stability within the structure of social environment. The generality of this particular exercise is also limited by both the low level of variance explanation and the highly specific and somewhat ambiguous nature of several of the factors. Given the above limitations this particular analysis was not pursued any further. It does, however, provide a useful framework within which to discuss the cross sectional factor analysis of urban social structure.

TABLE 5.2

Nottingham C. B.: Longitudinal Analysis of Social Area Stability: Factor Pattern Matrix

VARIABLE	FACTOR 1	2	3	4	5	6	7	8
1. Total Population	-0.4	0.02	-.001	0.03	-0.08	0.69	-0.26	-0.06
2. Owner Occupiers	-0.25	0.09	-0.78	-0.08	0.03	-0.04	-0.08	0.004
3. Local Authority renters	-0.38	0.03	0.73	-0.02	0.01	-0.11	0.04	0.08
4. Private, Unfurnished	0.74	-0.80	-0.03	0.16	0.05	0.14	-0.09	-0.14
5. Private, Furnished	-0.03	-0.07	0.06	-0.04	0.001	0.10	0.02	0.84
6. Share/Lack Hot Water	0.54	-0.05	-0.06	-0.44	0.23	0.006	0.15	0.20
7. Share/Lack Inside W.C.	0.38	0.07	0.28	-0.14	0.13	-0.001	0.61	0.006
8. Share/Lack Fixed Bath	0.75	0.03	0.07	-0.002	-0.12	-0.03	-0.27	0.02
9. Share Dwellings	0.11	0.07	0.04	0.72	0.17	-0.13	-0.08	0.36
10. Over-crowding	0.02	-0.005	-0.07	0.71	-0.10	0.18	0.15	-0.28
11. Coloured Immigrants	-0.03	-0.07	0.09	0.05	0.03	-0.07	0.82	-0.09
12. Population 0-14	-0.15	-0.07	-0.15	0.08	-0.22	-0.11	0.65	0.38
13. Population 65+	0.007	0.06	0.13	0.01	0.74	-0.11	0.15	0.06
14. 1 Person Households	0.15	0.01	-0.04	0.01	0.02	0.78	0.11	0.15
15. 6+ Person Households	0.01	0.20	0.22	0.01	-0.69	-0.17	0.15	0.08
16. Population 20-44	0.04	0.80	-0.10	0.15	-0.05	0.14	-0.24	-0.08
17. Population 45-60	0.12	-0.84	-0.03	0.06	0.03	0.07	-0.1	-0.09
18. White Immigrants	0.27	0.12	0.20	-0.01	-0.19	0.07	0.15	0.07

Social change: a cross sectional approach

Johnston (1973) suggested that a cross sectional factorial ecology was a more appropriate method for studying social change. By employing this methodology it is possible to pose two fundamental questions concerning urban social structure and its change/stability over time. Firstly, are the dimensions which segregate the study area the same in both 1961 and 1971? This is investigated by conducting separate analyses on the two data sets and comparing the resultant factor structures. Assuming that the residential structure is stable, it is then possible to consider if the intra-urban distributional pattern of these structures has been modified. This is reviewed by comparing scores on the relevant factor in each E.D.

Application of the same analytical procedure as used for the longitudinal investigation to the 1961 data matrix produced six factors with eigenvalues over one. These represented 76.8 per cent of the initial variation. For the 1971 analysis four factors with eigenvalues over one representing 83 per cent of the initial variation, were identified.

The factor pattern matrices resultant from these two solutions are shown in Tables 5.3 and 5.4. Interpretation of the 1961 loadings matrix suggested the following basis to the six dimensions:

(1) An urbanism factor in which shared dwellings, housing quality, elderly persons and ethnic minority variables are contrasted with the familistic measures of young adults and young children.

(2) A tenure measure which has positive loadings with the Local Authority tenure and large household indices and negative relationships with the owner occupier, private rented (unfurnished) and amenity provision measures.

(3) A deprivation index which links lack of housing amenities with overcrowding, small dwellings, unfurnished rented accommodation, large households and the yound dependent population.

(4) A private renting index.

(5) A housing quality factor which is described by the non-exclusive use of a fixed bath and hot water system measures.

TABLE 5.3

Nottingham C.B.: Factorial Ecology, 1961: Factor Pattern Matrix

VARIABLE	FACTOR 1	2	3	4	5	6
1. Owner occupiers	-0.08	-0.82	-0.26	0.08	-0.03	0.15
2. Local Authority renters	-0.01	0.89	-0.04	0.05	0.06	0.29
3. Rented unfurnished	-0.15	-0.33	0.38	0.55	-0.26	-0.10
4. Rented furnished	0.20	0.07	-0.08	0.87	0.21	0.17
5. Share/lack hot tap	-0.34	0.22	0.26	-0.27	0.33	-0.19
6. Share/lack fixed bath	-0.06	0.12	0.37	-0.04	0.86	0.12
7. Share/lack inside W.C.	-0.35	-0.33	-0.06	-0.02	0.26	0.05
8. 1-3 rooms	0.16	0.26	0.81	-0.04	-0.08	-0.21
9. Shared dwellings	-0.72	0.13	0.19	-0.02	-0.19	-0.05
10. 1 person household	-0.01	0.01	0.08	-0.22	-0.02	0.60
11. 6 person household	-0.05	0.41	0.33	0.05	0.14	0.50
12. Over-crowding	-0.07	0.27	0.67	-0.18	-0.16	-0.32
13. Coloured immigrants	-0.91	0.03	-0.02	0.07	0.03	-0.02
14. Population 0-14	0.72	0.03	0.49	0.23	0.06	-0.08
15. Population 65+	-0.89	-0.02	-0.06	0.17	-0.02	-0.05
16. Population 20-44	0.55	-0.28	-0.06	-0.17	-0.01	0.23
17. Population 45-60	-0.26	-0.07	0.03	0.01	0.06	-0.03
18. White immigrants	-0.31	-0.09	0.10	-0.05	0.07	0.10

(6) Household structure measure.

TABLE 5.4
Nottingham C.B.: Factorial Ecology, 1971:
Factor Pattern Loadings Matrix

VARIABLE	FACTOR 1	2	3	4
1. Owner occupiers	-0.03	0.22	-0.03	-0.87
2. Local Authority renters	-0.58	0.15	-0.32	0.69
3. Rented unfurnished	0.92	0.04	0.15	-0.17
4. Rented furnished	0.15	-0.02	0.94	-0.16
5. Share/lack hot tap	0.91	-0.17	0.29	0.03
6. Share/lack fixed bath	0.95	-0.13	0.19	0.04
7. Share/lack inside WC.	0.97	-0.05	-0.13	0.05
8. 1-3 rooms	-0.005	-0.08	0.87	0.08
9. Shared dwellings	0.18	-0.11	0.87	-0.09
10. 1 person household	0.12	0.44	0.73	-0.02
11. 6 person household	0.17	-0.73	0.11	0.39
12. Over-crowding	0.46	-0.41	0.52	0.19
13. Coloured immigrants	0.39	-0.24	0.64	0.04
14. Population 0-14	0.30	-0.85	-0.12	0.15
15. Population 65+	0.17	0.86	-0.13	0.04
16. Population 20-44	0.04	-0.44	-0.12	-0.52
17. Population 45-60	0.29	0.73	-0.12	0.13
18. White immigrants	-0.06	-0.16	0.32	-0.02

In 1961 therefore the urban social environment of Nottingham C.B. could be described in terms of 'three basic structures'. These constructs identified housing tenure (factors 2 and 4), housing quality (factors 3 and 5), and family status (factors 1 and 6). Little correspondence is demonstrated between these dimensions and the classic social area model postulates.

The four retained factors identified in 1971 were briefly described as :

(1) A housing quality index which contrasts low quality housing, overcrowding and ethnic minorities with Local Authority housing.

(2) A life cycle measure which contracts variables indicative of newly established families with single person households and the elderly and middle aged populations.

(3) An ethnicity measure which illustrates positive loadings with the ethnic minority, small dwellings, shared dwellings and private rented (furnished) variables.

(4) A tenure-life cycle measure in which large households and Local Authority tenure are contrasted with home ownership and the young adult population.

Thus the four basic axes of social segregation in 1971 may be summarised as housing quality, life cycle status, ethnicity and tenure-life cycle status. Whilst these constructs do not neatly correspond with the classic postulates of the urban social structure models, there is some evidence to suggest that these dimensions have moved closer to this residential mosaic typology than those identified in 1961.

The Dimensions of Social Segregation: A Comparative Analysis

Having identified the structure of residential segregation in the study area in both 1961 and 1971, the two factor solutions are now compared and their degree of similarity identified. Comparison of these factor structures may be achieved by three basic methods, namely - (1) comparison of loading matrices; (2) comparison of factor labels; and (3) statistical analysis. Each of these three methods is described below:

Timms (1970a) and Hunter (1971) both based their factor structure comparisons upon visual inspection of the relevant factor loading matrices. Considering the highly statistically sophisticated nature of the factoring procedure, such a technique is rather crude. It is also a highly subjective means of assessing the similarity of two (or more) factor loading matrices as the technique has no precise numerical quality. Given these limitations, however, cursory examination of the factor pattern matrices for 1961 and 1971 does suggest that the dimensions of differentiation were subject to considerable modification and redefinition during the decade.

An alternative procedure is to compare the nature of the factor structures as identified by the labels given to each factor. With the limitations inherent in the practice of factor labelling (Palm and Caruso, 1972) this procedure also has its drawbacks. Table 5.5 provides this information for the present study. Even with a crude analytical procedure it is readily apparent that there has been an underlying change in the structure of residential segregation in Nottingham C.B. between 1961 and 1971.

TABLE 5.5
Nottingham C.B.: Comparative Factorial Ecology 1961 - 1971

FACTOR	1961 label	1971 label
1	Urbanism	Housing quality
2	Tenure	Life cycle status
3	Housing quality - tenure	Ethnicity
4	Private rented tenure	Tenure - life cycle status
5	Housing quality	
6	Household structure	

Four of the six factors identified in 1961 identified housing, in terms of either tenure or amenity provision, as the key facet of social segregation. By 1971 the situation had considerably modified with housing no longer the dominant feature of the social environment as more subtle dimensions (e.g., life cycle and ethnic status) have emerged as equally important dimensions of residential segregation.

A more satisfactory method of assessing the precise degree of inter-relationship between the two factor structures is via the use of congruence coefficients. This technique was originally developed by Veldman (1967) and has several previous geographical applications (Johnston, 1973; Giggs and Mather, 1975). This particular procedure superimposes the two pattern matrices and one is then rotated to a position of best fit with the other. The greater the amount of rotation required the less alike are the two structures. By measuring the cosine of the angle between the two factor structures, it is possible to derive a numerical index which represents the similarity of the two solutions.

Table 5.6 represents the congruence coefficients resultant from superimposing the 1971 matrix on the 1961 solution. It is obvious from this table that the two solutions are extremely dissimilar. Giggs and Mather (1975) developed a simple descriptive scale for categorising the results of congruence analysis. This typology is: very clsoe fit (0.9 - 1.0), close fit (0.8 - 0.89), poor fit (0.70 - 0.79) and no fit (0.7). None of the coefficients in this particular study even attain the poor fit category. This evidence confirms the impression that the structure of residential segregation in Nottingham between 1961 and 1971 was subject to extreme modification.

TABLE 5.6
Factorial Ecology, 1961-1971: Congruence Coefficients

1961	1971 FACTOR 1	FACTOR 2	FACTOR 3	FACTOR 4
FACTOR 1	0.3046	0.4486	0.2998	0.0598
FACTOR 2	-0.4251	-0.2716	-0.1357	0.6306
FACTOR 3	0.3298	-0.4859	0.5165	0.1041
FACTOR 4	0.4844	-0.0593	0.5633	0.0258
FACTOR 5	0.1973	0.2176	0.3197	0.0617
FACTOR 6	0.2261	-0.0196	0.1769	0.1103

Discussion

This temporal investigation of the structure of the social environment of Nottingham C.B. has clearly identified that change over time is an important attribute of urban social structure. Both the longitudinal and comparitive factor analyses identified that the period 1961-1971 was characterised by a marked change in the structure of residential segregation in the study area. Clearly the intra-urban social fabric is a dynamic rather than static structure.

It is, however, important to note that this investigation did not validate the propositions of the urban change models which suggested that social, ethnic and family status would be the major axes of residential structure change. Rather this research revealed that a fundamental restructuring of residential segregation took place in the study area between 1961 and 1971. Between these two dates housing status, as defined in terms of both amenity provision levels and tenure, declined from its position as the prime axes of social segregation. Concurrent with this, specifically social dimensions emerged as social segregation factors of greatly enhanced power. Clearly, therefore, this decade was characterised by a realignment of the relationship between social and physical space in terms of the wider social environment.

In seeking to explain these identified changes in the structure of residential segregation, it is necessary to examine the post war planning policy of the County Borough. During the period 1954 to 1974, 18,661 slum houses were demolished with the bulk of this clearance occurring in the mid 1960's. This large scale clearance of slum properties in the city led to erosion of the extreme housing differentials descriptive of the study area in 1961. In turn this resulted in the demise of housing status as the dominant aspect of intra-urban social segregation in Nottingham. In response to, and in conjunction with this trend, more explicitly "social" (as opposed to the "physical" nature of housing indices) attributes such as family and ethnic status have emerged as more important social differentiators. This evidence therefore suggests that with the lessening of spatial inequalities in housing standards, social dimensions have been invoked as more pertinent axes of socio-spatial segregation.

This obvious change in the structure of residential segregation reflects the change displayed by individual elements of the social environment during this decade. Within the wider Greater Nottingham urban area (i.e., Nottingham C.B. and the five contiguous Urban Districts of Arnold, Beeston and Stapleford, Carlton, Hucknall and West Bridgford), it was established that population growth was a feature of the Urban Ring and decrease a characteristic of the city between 1961 and 1971 (Victor, 1979). The same work also established that, as well as changes in the overall size of the population during this period, modifications to the structure of the population were also evident. Out migration from the city to the urban ring was selective in terms of age, life cycle status, ethnic origins and social status. Housing development was also spatially segregated with private house construction centred in the urban ring and public development concentrated in the city. During this decade therefore, an important dichotomy in terms of overall social structure developed between the city and its surrounding urban belt.

A low status, elderly population, characterised by public housing, ethnic minorities and families in the latter phases of the life cycle became increasingly characteristic of the County Borough during the study period. Within the city, the overall urban area trends were accentuated with population decline at its most extreme in the inner city redevelopment areas. Victor (1979) clearly established the devastating effect which the excessive urban renewal schemes had upon the social structure of Nottingham in general and the inner city in particular. Against such a background, therefore, it is hardly surprising that the nature of residential segregation in Nottingham was modified between 1961 and 1971. As the actual residential axes were not stable over time it was not possible to review the stability of the spatial patterns.

As the introduction noted, little work has been published in this particular field of urban studies. There is, therefore, little with which to compare this study. Brown and Horton's (1970) work stressed social, ethnic and family status as the basic dimensions of change in Chicago. In the Toronto (Murdie, 1969) and Melbourne (Johnston, 1973) examples, stability rather than change in the urban structure was identified.

In view of the problems involved in cross-national comparisons of results, the only study with which these results may be meaningfully compared is the work of Davidson (1975) in Hull. Despite not having a constant set of spatial units or variables he compared, visually, the results of two separate Prinicipal Components Analyses produced for 1961 and 1971 data. He notes that there was a remarkable degree of similarity expressed by the two solutions. Such a situation is indicative of stability in the structure of the social enviornment. Social class, overcrowding, tenure and housing quality were the variables diagnostic of the two leading components for each date. Components 3 to 5 in both cases measured life cycle status families, female employment and nobility zone in transition elements of socio-spatial structure. Indeed, it was only with the sixth and subsequent components that there was any marked degree of change in structure. Clearly these results contrast markedly with the Nottingham situation. Given the methodological shortcomings of Davidson's work, however, it is difficult to draw a direct parallel with the Nottingham work. Further work is, therefore, required to establish whether the situation of extreme change in the structure of residential segregation, identified in Nottingham, is typical of the British urban context.

Conclusion

This paper has investigated two methods of identifying the major processes and patterns of social area change for 140 enumeration districts in Nottingham C.B. between 1961 and 1971. The first of these analytical procedures suggested by Brown and Horton (1970) was found to produce eight indices of change. However, on both methodological and conceptual grounds, it was felt that this procedure was inappropriate for the stated research objectives. The second method, a cross-sectional factorial ecology, illustrated the dynamic nature of the residential structure in a temporal context. The key feature of this trend was the enhancement of social features such as family and ethnic status as socially differentiating criteria. Housing policy initiatives by the Local Authority was the mechanism suggested as influencing this change in residential structure.

Johnston (1976) in this review of temporal studies of social area change, suggested that two main trends may be observed. Initially he proposes that the basic dimensions of social differentiation have remained stable over time. His second postulate suggests that there has been little change over time in the characteristics of most social areas in cities with the exception of areas experiencing rapid change in terms of their ethnic status. It is difficult to find supportive evidence for either of these generalisations in the present study. However, further work is required to establish if the conclusions described above are specific to the study area or are meaningful in a wider British context.

REFERENCES

Barke, M. Social deprivation in West Newcastle - some implications for area policies. In Fullerton, B. (ed.) North Eastern Studies Department of Geography, University of Newcastle upon Tyne (1977)..

Brown, A. and Horton, F.E. Social area change: an empirical analysis. Urban Studies, 7, (1970) 271-288.

Burgess, E.W., The growth of the city: an introduction to a research project. Publications of the American Sociological Society, 18,(1924)85-97.

Curson, P.M. The changing demographic structure of Auckland in J.S. Whitelaw (ed.) **Auckland in Ferment.** New Zealand Geographical Society, (1967) 23-39.

Davidson, R.N. The analysis of urban sub areas. S.S.R.C. project report, Department of Geography, University of Hull (1975).

_______ Social deprivation: an analysis of inter-censul changes. Transactions Institute of British Geographers, New Series 1, (1977) 108-117.

Davies, W.K.D. and Lewis, G. Urban dimensions in Leicester in Clarke, B. and Gleave, B. (eds). Social patterns in cities, special publication no. 5. Institute of British Geographers, (1973) 71-86.

Duncan, O. and Duncan B. The Negro population of Chicago, Universityof Chicago Press (1957).

Giggs, J.A. and Mather, P.M. Factorial ecology and factor invariance: an investigation. **Economic Geography,** 51(4), (1975) 366-382.

Hoyt, H. The structure and growth of residential neighbourhoods in American cities. Federal Housing Administration, Washington D.C. (1939).

Hunter, A.A. Factorial ecology - a critique and some suggestions. **Demography**, 9, (1971).

Johnston, R.J. Social area change in Melbourne, 1961-1966: A sample exploration. **Australian Geographical Studies**, 11 (1973) 79-98.

Johnston, R.J. Residential area characteristics: research methods for identifying sub-areas—social area analysis and factorial ecology, in Herbert, D.T. and Johnston, R.J. (eds.) Social areas in cities, volume 1, John Wiley, (1976) 193-236.

Morrill, R.L. The Negro ghetto: problems and alternatives. Geographical Review, 55, (1965) 339-361.

Murdie, R.A. Factorial ecology of metropolitan Toronto, University of Chicago, Department of Geography, Research paper 116 (1969).

Palm, R. and Caruso, D. Labelling in factorial ecology. **Annals of the Association of American Geographers**, 62, (1972) 122-133.

Robson, B.T. Urban social areas. Oxford University Press (1975).

Shevky, E. and Bell, W. Social area analysis. Stanford University Press, Stanford, California (1955).

Sweetser, F.L. Factor structure and ecological structure. **Acta Sociologica**, 8, (1965) 205-225.

Timms, D.W.G. Comparative factorial ecology: some New Zealand examples, **Environment and Planning**, 2,(1970a) 455-468.

________The urban mosaic. Cambridge University Press (1970 b).

Veldman, D.J. Fortran programming for the behavioural sciences. Holt, Reinehart and Winstan, New York (1967).

Victor, C.R. Social structure and social malaise: a geography of Greater Nottingham. Unpublished M.Phil. thesis, University of Nottingham (1979).

SIX

Gabriele Zanetto

Factors and Forms of Social Areas in a Conurbation

Since the Chicago sociological school proposed its first interpretation of the urban internal differentiation, studies on social areas within the city multiplied, accumulating empirical evidences as well as theoretical propositions. As the complexity of forces acting in this field becomes more evident, the possibility of a theory explaining the distribution of social, ethnic and economic characters of inhabitants becomes more remote (Timms, 1971). However, factorial ecology has surely allowed to test some hypotheses advanced about the relevance of some factors in the making of urban social areas; some typologies are consistently shown by factor loadings and scores, even if spatial patterns are widely different (Berry & Kasarda, 1977).

In general we can point out as relevant factors: (i) the dwellings availability, which is responsible for the "young" character of the most recently built areas, contrasting with an older population in settled areas, (ii) the accessibility to social services and commercial facilities, generally located in a central district, which determines, by means of land value, an opposition between centre and periphery or - more generally between areas of different value, (iii) a cultural and economical trend (for both individuals and public authorities) to set up homogeneous areas, by complex mechanisms of personal and political decisions - in such a trend we can find place for all ethnic, linguistic, religious and cultural differentiations. Undoubtedly, a considerable progress in the interpretation of urban mechanisms would be expected from a system approach (Beaujeu-garnier, 1980) able to get over both a determinist simplism and a behaviouralist indeterminateness.

A less inquired case is the social areas pattern with multiple nuclei. This is found in new conurbations formed by concomitant growth of two or more near cities, whose neighbourhoods come to touch each other. The involvement of region-wide areas in urbanization processes makes the analysis of a single city (whose borders are ill-defined but whose functions are well-evident) more and more hazardous.

Such a problem is particularly evident in the case of the conurbation of Venice (400,000 inhabitants), formed of two main centres: the first in a lagoon (the historic Venice: 100,000 inhabitants), the second in the mainland (Mestre: 150,000 inhabitants), both surrounded by some minor centres and linked through a ten kilometres bridge over the lagoon. Even if deep functional relations are well established within the whole area (Zanetto, 1982), it is not clear whether the whole lagoon side and the mainland side of the conurbation are socially integrated as parts of a unique system or, on the contrary, they are two parallel cities - even if socially differentiated (Lando & Zanetto, 1980).

In order to answer such a question a principal component analysis has been carried out, over 32 variables and the 344 census areas of the commune of Venice in 1971. Variables are related to demographic characters of residents, their professional status, their sector of activity, their education level, the quality of dwellings. Variables have been standardized before the analysis, and factors have been submitted to a varimax rotation (further details in Zanetto, 1981). No significant ethnic differentiations is found in the area.

The analysis points out five relevant factors, even if only three are responsible for nearly two-thirds of total variance (24.2; 22.5; 15.8 in the order; eigenvalues are respectively 8.2; 7.6; 5.4), while the fourth and fifth factors are each strongly related to a unique variable - and their scores identify small exceptional areas, such as military or rural settlements.

The first component is the classical differentiation of socio-economic status, as it opposes professionals, entrepreneurs, managers and employees to blue-collars; workers in banking, insurance and service industries to workers in manufacturing and transportation, well-educated to non-educated people; large dwellings to small and crowded ones. So the conurbation of Venice experiences as the majority of cities - a dominant differentiation between "rich" and "poor" areas, characterized, moreover, by an age element: factor loadings show that high status areas are also older than low status areas.

The first component scores are distributed very regularly in the urban space, around the three highest areas coinciding with two Central Business Districts (in Venice and Mestre) and with the core of the island of Lido; a green residential area between the lagoon and the sea. The more we move away from these centres, the lower the status becomes, with the exception of the extreme periphery where it rises again. As a whole, the mainland side has more low status areas than the lagoon side. This can be explained by a remarkable concertration of work places in the historical centre, in spite of its lower population; the higher cost of dwellings we find there confirms that an income selection is operated in its population. Nevertheless the conurbation is shown by the fisrt component as a sum of two twin-cities, and a study on the perception of the quality of residences in the same area (Balboni, et al., 1978) points out such a symmetry as a dominant component of desiderability; high and low scores were charged both to lagoon and mainland areas.

But the apparent denying of an organic link between the two, as proposed by the first component, is contrasted by the scores of the second component, which recognizes differences between lagoon and mainland. The second component opposes (as the first does, but more strongly) a younger area (with large families, workers in manufacture) to an aged area, characterized by tertiary activities of its residents. But, rather than adding to the old/young opposition a status feature, the second component qualifies the aged zone as an olderly urbanized one, with high density of population and fewer home-owners. It defines the old town, quite exclusively localized in the historic centre in spite of its various status, even if highest scores avoid the high-status core. In the "younger" area, both people and urbanization are younger (recent buildings, low density, ownership, no gaz network), and we can find some surviving agricultural and fishing activities, as well as construction workers.

The area so defined covers a large part of the mainland and some minor islands, but highest scores are concentrated in the furthest periphery of both sides of the conurbation. So this component seems to denote a strong complementarity between them, defining the mainland as an enlargement of the lagoon town, even if both have an internal structure defined by the status. It is interesting to note that such an opposition between lagoon and mainland is found as a prinicpal component of the above mentioned residential desiderability. Hence we could suppose that the opposition between the old and the new city has even some cultural roots. As the matter of fact, everyday life is very peculiar in Venice, given the particular feature of the city and the lack of traffic circulation, which is easily acceptable only to natives.

Our third social component of spatial differentiation, with its few exceptional scores concentrated in small areas, and its loadings on variables defining a very low quality of dwellings and analphabetism, points out the severe poorness of some ghettos near the industrial port and in the furthest periphery.

The factor ecology of the conurbation is therefore based on the traditional opposition high vs low status and new vs old city, as we could expect (Foggin & Polese,1976). The perception of the inhabitants confirms the existence of two twin-cities from a status point of view, and of a unique city from a cultural point of view. So we have tried to draw a general regionalization, disponing the 344 areas on the space determined by the two principal components. Their position present a sharp gathering of four dominant groups, over the four half-axis. A discriminant analysis has defined the four groups shown in Fig. 6.1 after three iterations (see Racine & Reymond, 1973).

Considering the geographical position of the areas so defined (with no contiguity constraints), we can find: (A) a central area, defined by high status and spatially segmented into three nuclei; there are the richest zones, near the largest concentrations of services, offices and business; (B) a blue-collar neighbourhoods, defined by low status, which covers a wide part of the town in the mainland; there is a very compact urbanization, relatively recent, of which we can find a little sample in Venice too; (C) the historical city, determined by age, consisting of old buildings, high density, aged and non-manufacturer population; there is the town of Venice, except its core which - because of the intense urban renewal it has experienced (see Costa, et al., 1980) - is a richer and newer neighbourhood; (D) the periphery, not yet completely urbanized, surrounding the whole conurbation both in the lagoon and in the mainland (see Fig. 6.2).

As we have seen, an opposition is established between (A) and (B) areas, as well as between (C) and (D), while in both couples we find a zone present in a unique side of the conurbation and another lying in both sides: lagoon and mainland. The blue-collar area has its counterpart in the high-status nuclei; the old area is balanced by a general periphery. This evidence lends support to the idea that we must consider the conurbation as a unique social system, and we can confirm this statement analysing the intra-urban migration progress, as suggested by Greer-wootten, 1972. This evidences a wide replacement in the historic Venice, producing a constant rise of status, aging, and a global decrease of its population (further details in Costa, et al., 1980). Such a phenomenon is mainly due to the expansion of

FIG. 6.1

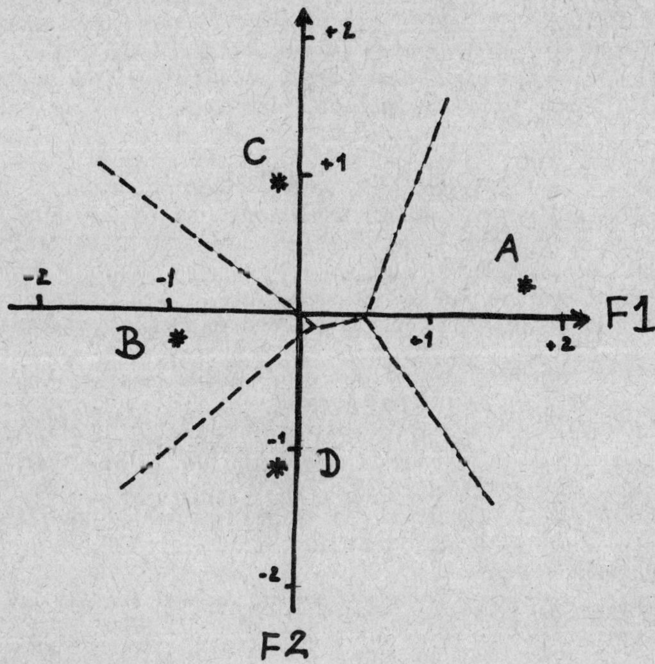
+2
C
+1
A
-2
-1
F1
B
+1
+2
-1
D
-2
F2

FIG. 6.2

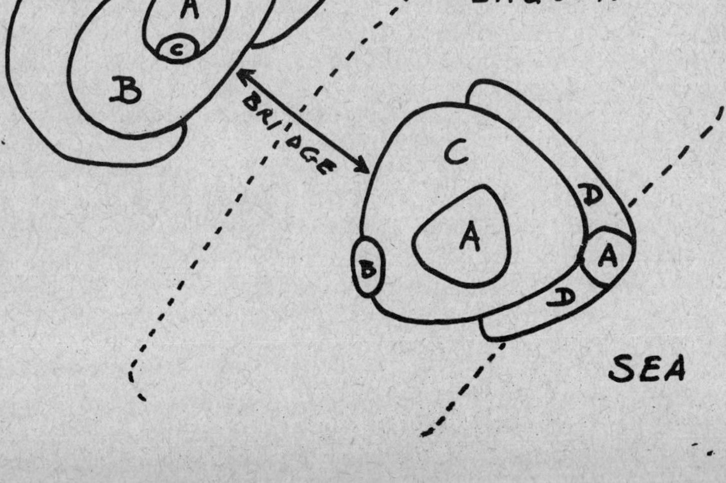
MAINLAND
D
D
A
C
B
LAGOON
BRIDGE
C
D
A
B
A
D
SEA

tourism and offices, which expels mainly middle class from the core of Venice. The older and blue-collar neighbourhoods, on the contrary, are more stable, as the urban growth has skipped them, turning to an outer periphery where we find the youngest city and a middle (and sometimes mixed) status.

The equilibrium of the conurbation is therefore based on two main residential areas surrounding a concentration of work places in manufacture (in the mainland) and in tertiary activities (in the historic city), both endowed with a central district and surrounded by a common periphery. Only a further expansion of tourism, or a severe decline in manufacture could alterate this equilibrium, moving toward a complete unification of the conurbation.

REFERENCES

P. Balboni et al.,**La percezione dell'ambiente. L'esperimento di Venezia,** Venice, Ciedart, (1978).

J. Beaujeu-Garnier, "Essai sur le systme urbain", **Acta Universitatis Carolinae, Geographica,** 15(1980), Supplementum, Praha, pp. 111-124.

B. Berry & J.Kasarda, **Contemporary urban ecology,** New York, Macmillan, (1977) (particularly pp. 108--157)

P. Costa et al., "Rinnovo urbano e trasformazioni sociali nel centro storico di Venezia", **Sistemi Urbani,** Naples," 2(1980), pp. 385-410.

P. Foggin & M. Polese, **La geographie sociale de Montreal en 1971,** "Etudes et Documents" n. 1), Montreal, I.N.R.S. Urbanisation, (1976).

B. Greer-Wootten, "Changing social areas and intra-urban migration process", **Revue de Geographie de Montreal,** 26(1972), pp. 271-292.

J.B. Racine & H. Rey Mond, **L'analyse quantitative en geographie,** Parts, Presses Universitaires de France, (1973).

D. Timms, **The urban mosaic. Toward a theory of residential differentiation.** Cambridge, University Press, (1971).

G. Zanetto, "A definition of Venice in a metropolitan context", in R. Bennett (ed.), **European progress in spatial analysis,** London, Pion, (1981), pp. 297-300.

ID., "Schema di interazione e compatibilita tra usi econmici di una zona umida", in M. PINNA (ed). **Salvaguardia di laghi e zone umide,** Rome Societa Geografica Italiana, (1983) (forthcoming).

ID. & F. Lando, "Mestre: analisi tipologica di una struttura urbana", **Bollettino della Societa Geografica Italiana,** 117(1980) Rome, pp. 213-255.

SEVEN

Gerald Walker

Social Structure of Neighbouring, Working People in West End, Toronto

SOCIAL NETWORKS, NEIGHBOURING AND NEIGHBOURHOOD

Three conceptions may be used to characterize the informal patterning of social interactions. Social network analysis deals with fairly stable sets of interaction intimates. These networks of friends and relatives have been found in formal organizations also, and, it has been argued, they are the building blocks of higher order power groupings.[1] Further, it is clear that social network structures are structurally diverse. Networks affect flow of information and patterns of diffusion of innovations. The social network tradition provides a powerful instrument for the study of micro-social geography.[2]

Neighbouring, as a concept, has been more idealogical assertion than useful social geographic concept. However, it seems clear that there is a level of neighbouring interactions in limited territories. The concept is useful because it does not restrict patterning of relations. Here it will refer to the drive for interactions among near dwellers.

Neighbourhood, by contrast, though redolent with idealogical overtones has gained some operational precision through empirical study.[3] Neighbourhoods come in a variety of shapes, but all of them are high order territorial structures. Suttles used the idea of ordered segmentation as the basis for examining neighbourhood in the slums of Chicago.[4] Under varying circumstances limited territories can become mobilized. They then generate local defense institutions and a sense of social and spatial identity emerges. Embattled upper and lower class residential areas, as defended neighbourhoods, have been studied. Further, neighbourhoods can gain a collective, and even corporate, character far short of highly mobilized, defended neighbourhoods.[5] (Figure 7.1)

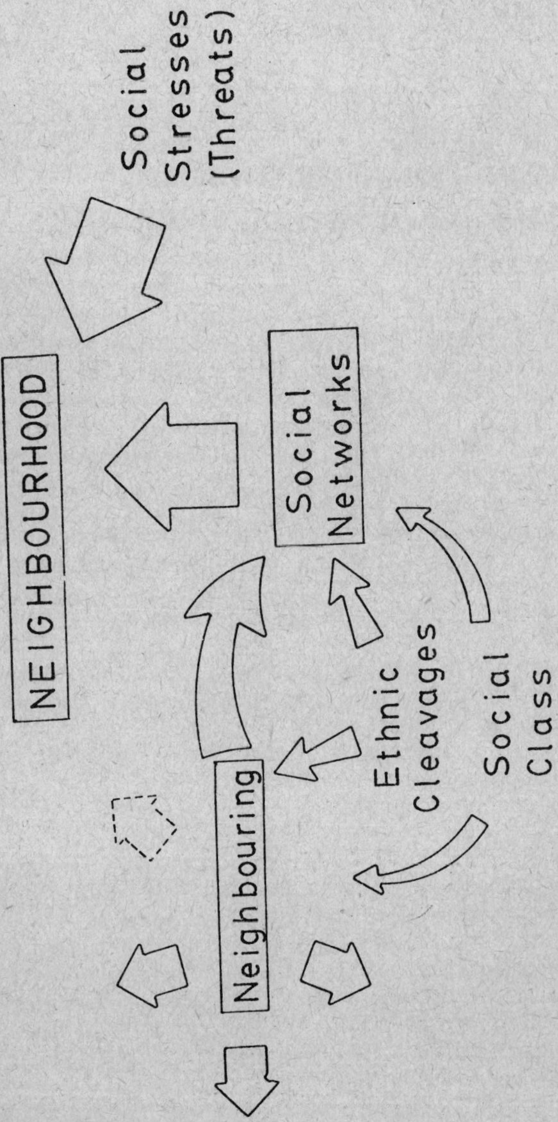

FIG. 7.1
Relationships in Local Territory

Between them, these three concepts provide a means of approach to the social structure of social interactions. This study is an attempt to use these ideas, especially the social network concept, to examine a part of the inner city. Several questions and hypotheses have guided the collection and examination of the data to be reported here. What are the sizes, structures and frequencies of interaction in a working class residential district? Is there evidence that neighbouring is important, either in itself, or as a basis for network construction, or in the creation of neighbourhood?

In a recent study Barry Wellman has argued that the metropolitan region has become more important than local neighbourhood in patterns of informal interaction in Toronto.[6] My own impression in several studies I have supervised is that while the entirety of Metro Toronto, and a large commuter shed beyond its municipal boundaries, is the key interaction zone, there is a lively continuing neighbourhood interaction system.[7] We appear, in fact, to have two elements at work in the structuring of informal social interactions in Toronto.

On the one side, the natural processes of spread of family and friendship groups is not impeded or seriously deflected by tracts of unsafe residential terrain. Toronto is a relatively safe and almost benign place to live. The active principles in residential selection are money, the desire to live in ethnic, life cycle and life style locations of preference. Avoidance through fear is a small aspect. Consequently, it would appear that local ethnic concentrations tend to disperse quite quickly and invasion-succession processes operate in very complex territorial mosaics. The other side of this structure of peaceful urban relations is that very few neighbourhoods have had to mobilize their human resources and become defended neighbourhoods.

The 1960's witnessed massive ethnic in-migration to Canada. Urban renewal occurred, at the expense of working class neighbourhoods, near cities' centres. The responses in Toronto were the appearance of a cosmopolitan city where a dour, north British, provincial city had been. Some neighbourhoods did mobilize and allied themselves to compete with the business establishment for power. A Conservative-Socialist political alliance emerged at city hall. Anti-development, pro-neighbourhood rhetoric became the order of the day by 1972. When the political situation stabilized, inner city development had been slowed and neighbourhoods no longer maintained their heightened defensive postures.[9]

This is not to suggest that Toronto is an urban idyl. The

insurgent alliance is gone. Developmet is on the move, and was never completely halted. Inflation, government wage restraints, increased energy bills and a host of other problems stemming from low incomes, make working class life in Toronto a struggle. However, the relatively pacific urban scene makes Toronto an easy, as well as interesting city to study.

Social networks have rural and urban forms.[10] I believe these are monentary crystallizations of two underlying modes of social and territorial organization. Social networks seem to be either largish, ramified, with loose structure, but with clear neighbourhood form: or alternatively, small, tightly knit and usually sub-neighbourhood in form. The first is typically rural, or appears in impacted urban situations, creating urban villages. The second is typically urban, most commonly found in residential areas with minimal social stress. I believe, further, that these two network structures are the outward manifestations of divergent tendencies in human organization. These tendencies stem from societies' division of labour and resultant forms of social solidarity.[11] In this framework neighbourhood first appears as high order structure, as diverse social networks are knit together, and institutionalized, in stressful situations.

Within this schema, neighbouring may be envisioned as a low level tendency to form bonds in local space. Social networks and neighbourhood formation require reinforcement from forms of social solidarity through division of labour and stressful conditions.

Given links between neighbouring, networks and neghbourhoods, and the relatively benign conditions of Toronto, what might we expect in a stable working class residential district? Without the need for a defended neighbourhood, the high order institutions would not be found and the local concept of neighbourhood would be vague, probably associated with traditional, named areas of the city. Spatial appraisal would be conditioned by city wide monetary and status evaluations. Neighbouring ought to continue, but local neighbouring ought to be a relatively weak part of interactions. Finally, social networks should serve largely personal and immediate functions. They would not become quasi-institutions, as in the gossip networks of the slums.[12]

We might, further, expect social networks to be formed on the basis of whatever social cleavages exist in the area. Still further, social networks could be the mechanism for establishing links between social groups where defense institutions are nonexistant and where hostile oppositions of ordered segmentation are not dominant urban motif. Thus, social networks without profound neighbourhood overlay would be expected to carry most of the interactional burden of the local area.

In Chicago, Kornblum has found that ethnic composition was critical in network formation. Some formed ethnically homogeneous networks while others formed ethnically heterogeneous networks, establishing links between groups.[13] Laumann has found, in Detroit, that nativeborn Americans still used ethno-religious structure as a primary basis for network formation.[14] It seems reasonable to expect to find ethnically based networks in Toronto, also.

Working People in West End, Social Composition

In order to test hypotheses and evaluate the larger schema, interviews were collected in a residential area of Toronto's West End. The West End is a sprawling, mixed land use area of Toronto. It includes much of the city's older industrial district. A fair portion is in institutional use, the rest is largely working class housing. Very little of West End has become fashionable. The upper middle class white painters, who have transformed Yorkville, Cabbage town and are working on High Park, are not yet in evidence.

The area is a welter of ethnic groups. Although Italians, Portuguese, Greeks, Ukrainians and Jews have been recognized as having their own territories, these high ethnic concentrations are rare nodes in a sea of mixed ethnic occupancy.[15] An ethnically mixed, but stable, working class area was sought for interviews. After a preliminary period of search, Census Tract 55 was chosen. The area was a small, stable working class district, south of Bloor, west of Ossington, east of Dovercourt and north of College (Fig.7.2). In 1971 it contained 1,105 occupied dwellings, the sample taken, of 96 households represented 8.7% of those dwellings.[16]

The sample was taken through a series of spatially random initial interviews, followed by chain referal interviews. When a given chain ended, a new random initiation point was selected. The aim was to combine a random sampling frame with a structured interview sequence which would allow the social structures of neighbouring to be uncovered. The interviewed population is broadly representative of the population at the 1971 census. English Canadians (British in the 1971 census) were underr presented. Catholics were over represented. However, these anomalies are accepted as representing the shifts in population composition between 1971 and 1977.

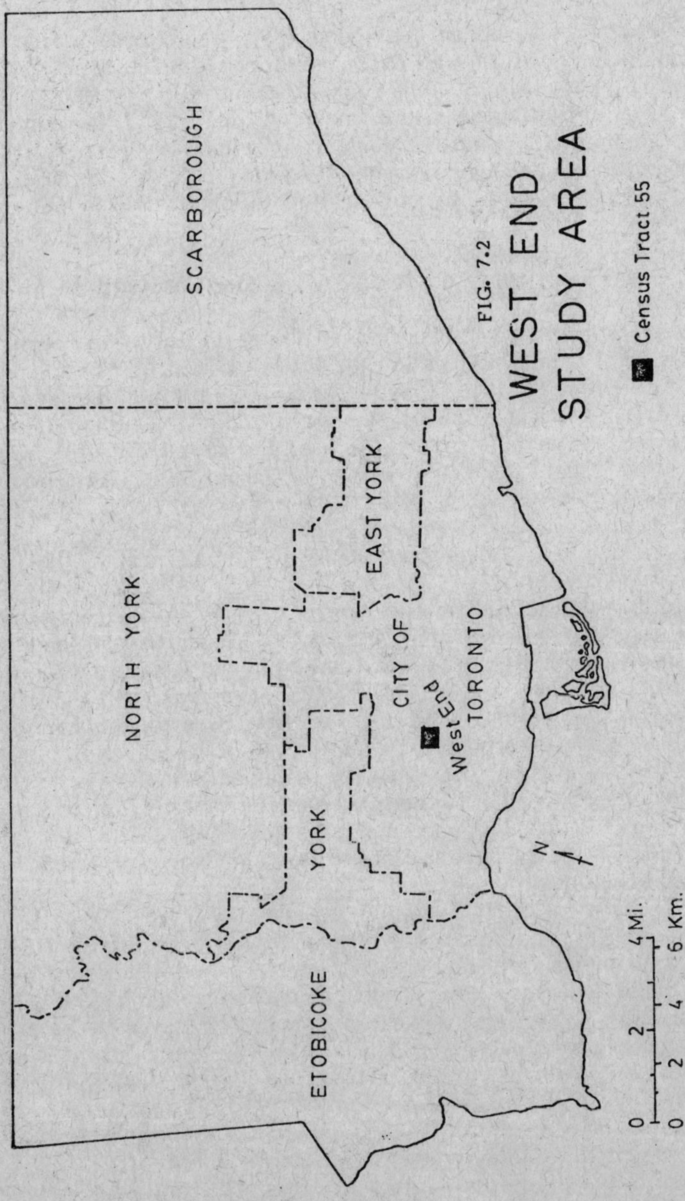

FIG. 7.2 WEST END STUDY AREA

The sample was a stable, ethnically diverse working class population (Table 7.1). The majority of the respondents (60%) were unequivocally blue collar working people. The majority of these were skilled workers, followed by semi-skilled and general labourers. About 15% of the sample could be defind as middle class, mostly lower middle class white collar workers. Over 6 per cent of the sample were unemployed working people, while another 9 per cent were retired working people and 2 per cent were students. The labour force participation rate was certainly as high as in the norm in Toronto. Unemployment was less than that of Toronto at the time of the interviews.

Incomes for families were much lower than the Toronto average. For Metro as a whole, family incomes were at about $17,000 in early 1977. The sample average was $10,000 for the 80 per cent of the sample which responded to the income question. This figure was quite close to that of a comparable working class sample I had collected in East York at about the same time. While the average incomes stood at about 60 per cent of the Metro average, the distribution suggested that incomes, while low, were adequate for household survival. About 25 per cent of the sample were poor. Another 25 per cent were modestly affluent. The remainder had low but adequate incomes. Further, home ownership stood at 73 per cent, much higher than in Metro as a whole. 35 per cent of the sample had completed high school. A relatively high 15 per cent had taken some post secondary education. Educational levels indicated a substantially literate population.

The West End sample was ethnically quite diverse. The two largest ethnic groups were Italians and Portuguese, English Canadians were third, Greeks fourth and Ukrainians comprised the fifth largest ethnic group. Geographically, ethnicity was distributed in a number of near neighbour pockets, with no portion of the study area dominated completely by any one ethnic group (Fig.7.3).

Religion was less complex, at least on the surface, than ethnicity. Catholicism was the majority faith among respondents. Protestantism was largely confined to English Canadians and a few Germans. Most Protestants were Anglicans or United Church members. Almost 18 per cent of the sample were Orthodox Christians. These were split between the Greek population, who were all Orthodox and the Ukrainians, half of whom were Orthodox, the remainder being Uniate Catholics. Jews were only 1 per cent and, with agnostics, other religions and no religion, made up about 10 per cent of the sample.

TABLE 7.1
Social Composition of Interviewed Population

Ethnicity

Portuguese	19	(19.8%)
Italian	17	(17.7%)
Ukrainian	8	(8.3%)
Polish	3	(3.1%)
Greek	11	(11.5%)
Chinese	3	(3.1%)
East Indian	4	(4.2%)
English Canadian	14	(14.6%)
Other European	14	(14.6%)

Religion

Catholic	53	(55.2%)
Orthodox	17	(17.7%)
Protestant	14	(14.6%)
Jewish	1	(1%)
Agnostic	1	(1%)
Other		
No Religion	3	(3.1%)
	5	(5.2%)

Occupations

Student	2	(2.1%)
Blue Collar Labour	57	(59.4%)
Prof/Manag	7	(7.3%)
Self-Employed	7	(7.3%)
Retired	9	(9.4%)
Unemployed	6	(6.3%)

Husband's Age

$\bar{X}$ = 40.3
S_X = 18.76

Children at home (under 16)

111 in 56 households
X = 1.98 S_X = 1.07

Length of Current Residence

$\bar{X}$ = 11.45 S_X = 14.39

1-5	= 44	(46%)
6-10	= 16	(17%)
11-20	= 19	(20%)
21+	= 17	(18)

Voluntary Organization Memberships (Local)

15 households in 1-3 organizations
X = 1.20 S_X = .56

Voluntary Organizations (Non-Local)

43 households in 1-5 organizations

$\bar{X}$ = 1.26 S_X = .79

Education of Male

0-8 years	33	(34.5%)
8-13 years	32	(34.5%)
13+ years	14	(14.5%)

Household Income

$ 0- 4,000	5	(5.2%)
$ 4,001- 8.000	18	(10.8%)
$ 8.001-12,000	26	(27.1%)
$12,001-16,000	25	(26 %)
$16,001-20,000	4	(4.1%)

$\bar{X}$ = $10,256 N = 78 (81%)

S_X = $4,043

Home Ownership

Owner	70	(72.9%)
Tenant	23	(24 %)

Marital Status

Married	56	(58.3%)
Single	29	(30.2%)
Widow/Widower	8	(8.3%)
Div/Sep.	2	(2.1%)

Household Numbers

$\bar{X}$ = 4.05 S_X = 2.05

Household with Roomers

35 (36%)

Residential Satisfaction

Low	= 1	=	3	(3.1%)
	2	=	1	(1%)
Average	= 3	=	58	(60.4%)
	4	=	12	(12.5%)
High	= 5	=	19	(19.8%)

X = 3.35 S_X = 1.11

Know Next Door Neighbour

78 (81.3%)

Neighbouring

been in house	51	(53.1%)
visit regularly	10	(10.4%)

Friends Predominantly Local

26 (27.1%)

Source: Field Interviews, 1977.

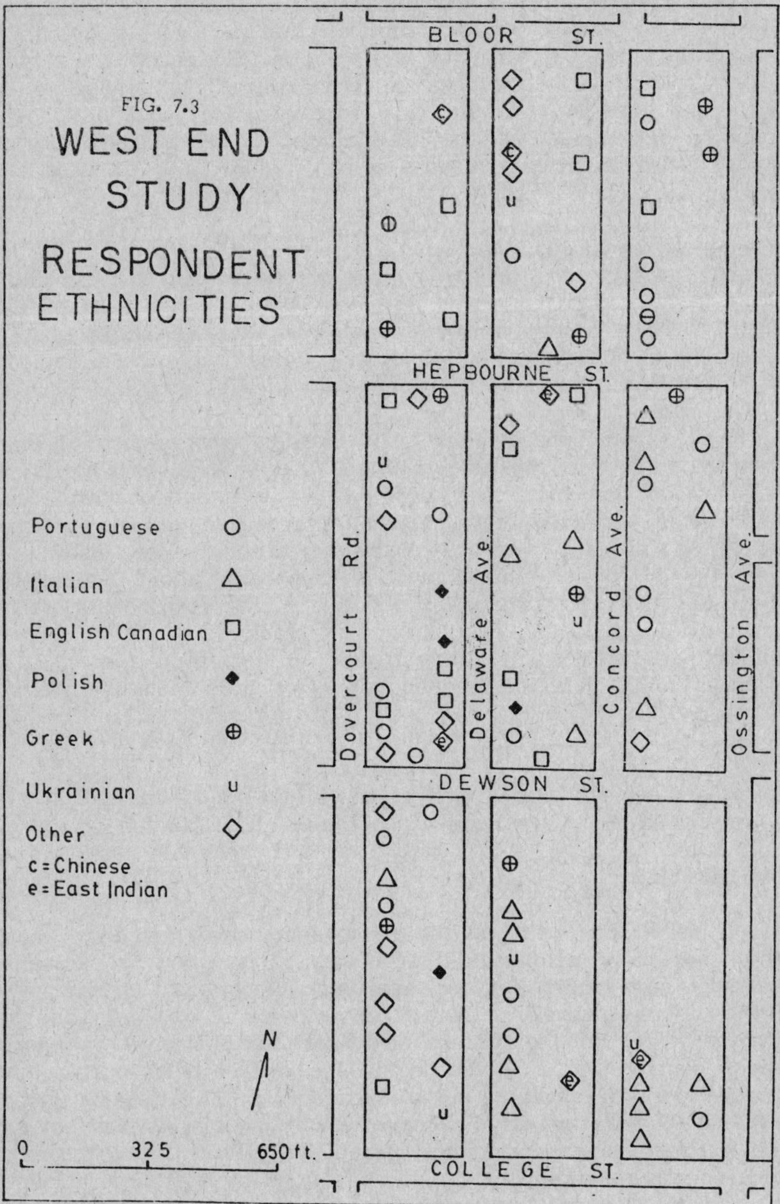
FIG. 7.3
WEST END
STUDY
RESPONDENT
ETHNICITIES
Portuguese
Italian
English Canadian
Polish
Greek
Ukrainian
u
Other
c = Chinese
e = East Indian
N
0
325
650 ft.
BLOOR ST.
HEPBOURNE ST.
DEWSON ST.
COLLEGE ST.
Dovercourt Rd.
Delaware Ave.
Concord Ave.
Ossington Ave.

When ethnicity and religion were cross tabulated four prominent ethno-religious groups emerged: Portuguese Catholics, Italian Catholics, Greek Orthodox and English Canadian Protestants. About 60 per cent of the respondents could be classified jointly by ethnicity and religion. Ukrainians were most likely to fail such grouping. Further, Ukrainians tended to be an older, residual population. However, for most purposes, religion and ethnicity were encapsulated in interactional groups and this combination presents a major structure of relations within the study area.

The interviewed population, mostly in their active working years, usually lived in two to six person households, typically with 2 or 3 children under 16. Only about 10 per cent of the sample lived in non-family households. Over a third of the respondent households had roomers living with their nuclear family.

The sample householders were largely home owners of long residence, both in the neighbourhood and in their own homes. A high proportion of the respondents were satisfied with their residences and with their residential area. A semantic diffierential and several questions concerning desire to live elsewhere were used. 75 per cent responded favourably about the neighbourhood district, although about 1/3 of the respondents would rather live elsewhere and almost half would move if the opportunity presented itself. Almost half of the population in the sample had lived in the neighbourhood for more than five years.

SOCIAL NETWORKS AND NEIGHBOURING

Our data permits examination of three aspects of informal social interactions among the sample population: diffuse neigh- bouring, formation of local social networks and voluntary association memberships.

Over 80 per cent of the respondents said that they know their next door neighbour, though only 53 per cent had actually been in their house. Furthermore, only 10 per cent of the respondents regularly visited their next door neighbours. Neighbouring, in the diffuse sense of associating with others living nearby, was at a low level. Yet, diffuse neighbouring was usually part of ramified network structure. The neighbourhood as largely made up of a group of known people: It was not an area where neighbours were strangers. Though neighbouring visiting was not dominant, it was practiced by most householders.

The structure of social networks adds another dimension to

neighbours. Each respondent was asked to provide information and locations of three intimates located in the study area. It seems that the census tract was the base for a complex set of local informal social interactions. Discussion of network findings will include characteristics of individual network embeddment and aggregate network structures.

Most of the respondents were part of local social networks, though about 1/3 were local social isolates. However, it would appear that many of the isolates and most diadic members were involved socially in the larger Metro setting. Most of the following discussion, however, will concentrate on the network characteristics and embeddment of the 2/3 of the respondents with local intimates (Table 7.2). The data was mapped and prepared in topological diagrams.

TABLE 7.2
Social Network Embeddment for Individuals

Network Variable	X	S_X	N
Times Chosen	1.39	.61	46
Times Choosing	1.74	.81	57
1st Order Star	3	1.09	65
Reciprocal Choices	1.26	.44	31
Total Linked Network	7.12	4.66	65
1st Order Zone (Links)	2.63	1.44	65
2nd Order Zone (Links)	5.18	3.26	65
SPAN*	.47	.28	65
DENSITY**	.75	.32	65
Ave Friend Dist (mi)	.36	.60	41
Ave Kin Dist (mi)	.53	.60	24
Frequency (monthly)	1.77	.86	65

Source: Field Interviews, 1977
* SPAN = E/S where E = 1st Order Links, S = Total Net Size
**DENSITY = $\frac{Na}{N(N-1)/2}$ where Na = interlinkages, N = 1st Order Stat.

Excluding local social isolates, respondents were part of networks of from three to eleven local households. The average of seven in a network comes quite close to the number (5.6) estimated by respondents for the size of their intimate social circle. A substantial portion of the choices were reciprocal. This is partly an artifact of the method of data collection. Chain referral interviews will clearly enhance these choices.

The respondents visited twice as many local friends as kin. However, both kin and friends were very numerous. The numbers of friends of friends, second order stars, was substantial, averaging two more than the first order core of intimates. It is this group which knit a number of sub-networks together. Typically our chains terminated before these peripheral people could be interviewed. It is likely that many more links existed than were recorded.

Interaction frequency at 1.77 meant that most respondents visited in the vicinity of from four to six times a month. This is a low figure, but is comparable to visiting frequencies in the countryside among both rural and exurban people.[17] The pattern of interlinkage among network members gives insight into the interaction patterns. Span, the proportion of direct contacts among the total linked group, was 47 per cent, a high figure, indicating that most chains were short and were made up of large numbers of people known directly to one another. Density, or the existing proportion of potential links between first order star members, was quite high at 75 per cent. This means that the networks, small to middle sized, were quite highly interlinked.

Distances of interaction showed an expected decay functions for friends and kin. Respondents were willing to travel almost twice as far to visit relatives as they were to visit friends (Table 7.2 and Fig. 7.9).

These embeddment statistics are quite similar to those found among suburbanites and dispersed exurbanites interviewed in the countryside, North of Toronto. While there were differences in collecting density statistics, density appears to be higher than among suburbanites.

Aggregate network statistics further illustrate social structure (Table 7.3). Most contacts were with friends, though a core of kin was found in many networks. Generally, interaction partners had been known for years. Relations were stable and long lasting. As well, the ordering of intimates was largely on the basis of time known, with oldest friends first named.

The social composition of the network associates provided an interesting contrast to the interviewees. English Canadian Protestants were significantly under represented. It is my impression that the English Canadians were less local in interaction than other ethnic groups. Portuguese and Italians were the most local in their orientations. While active working people were the largest group in occupational terms, they were still proportionately underrepresented. Also, students, retired people, unemployed and self-employed were more locally oriented groups.

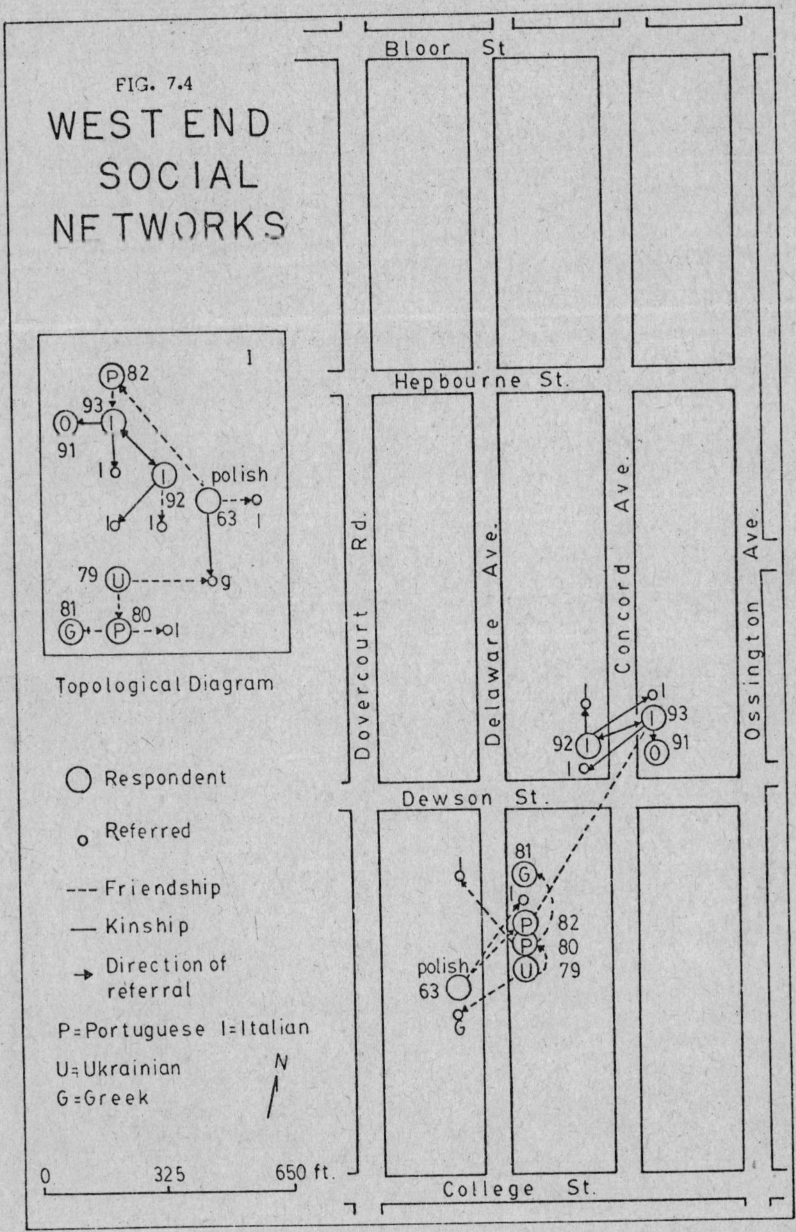
FIG. 7.4
WEST END SOCIAL NETWORKS
Bloor St
Hepbourne St.
Dewson St.
College St.
Dovercourt Rd.
Delaware Ave.
Concord Ave.
Ossington Ave.
Topological Diagram
Respondent
Referred
Friendship
Kinship
Direction of referral
P=Portuguese I=Italian
U=Ukrainian
G=Greek
N
0
325
650 ft.
polish
63

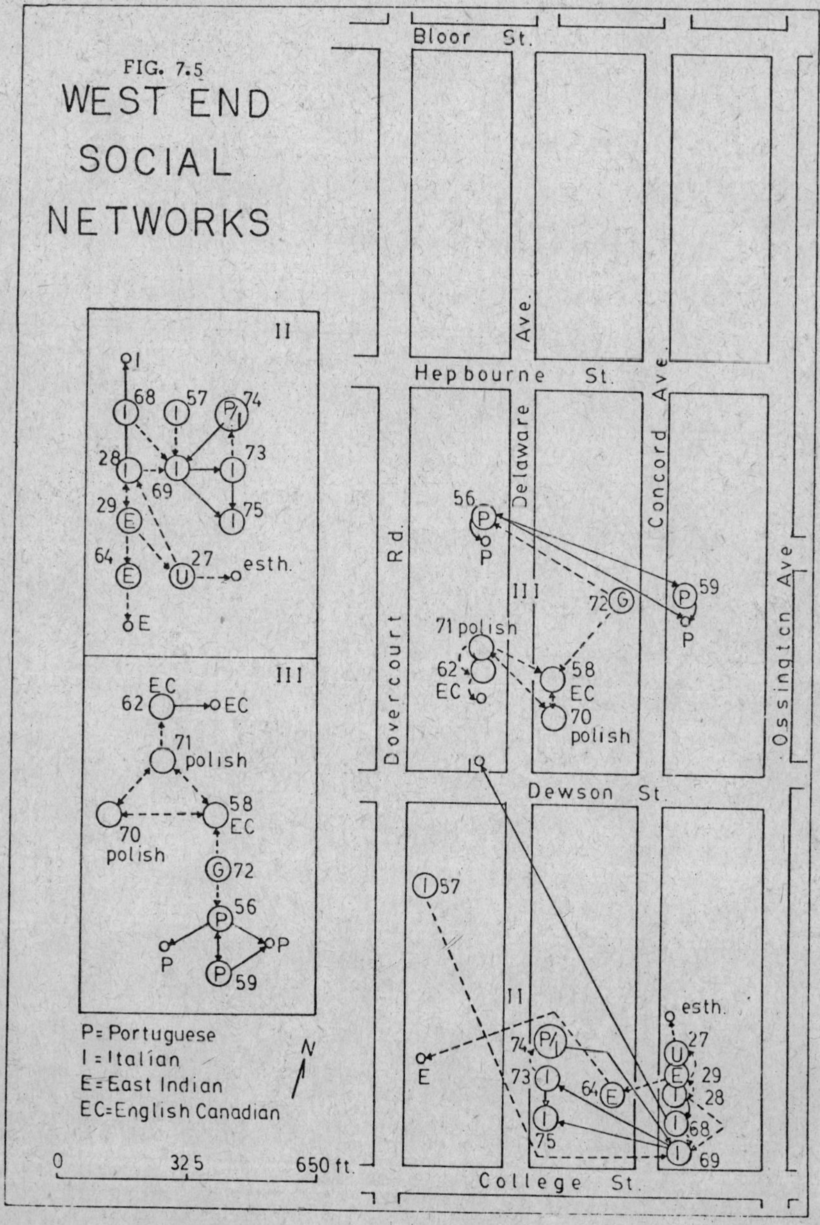

FIG. 7.5 WEST END SOCIAL NETWORKS

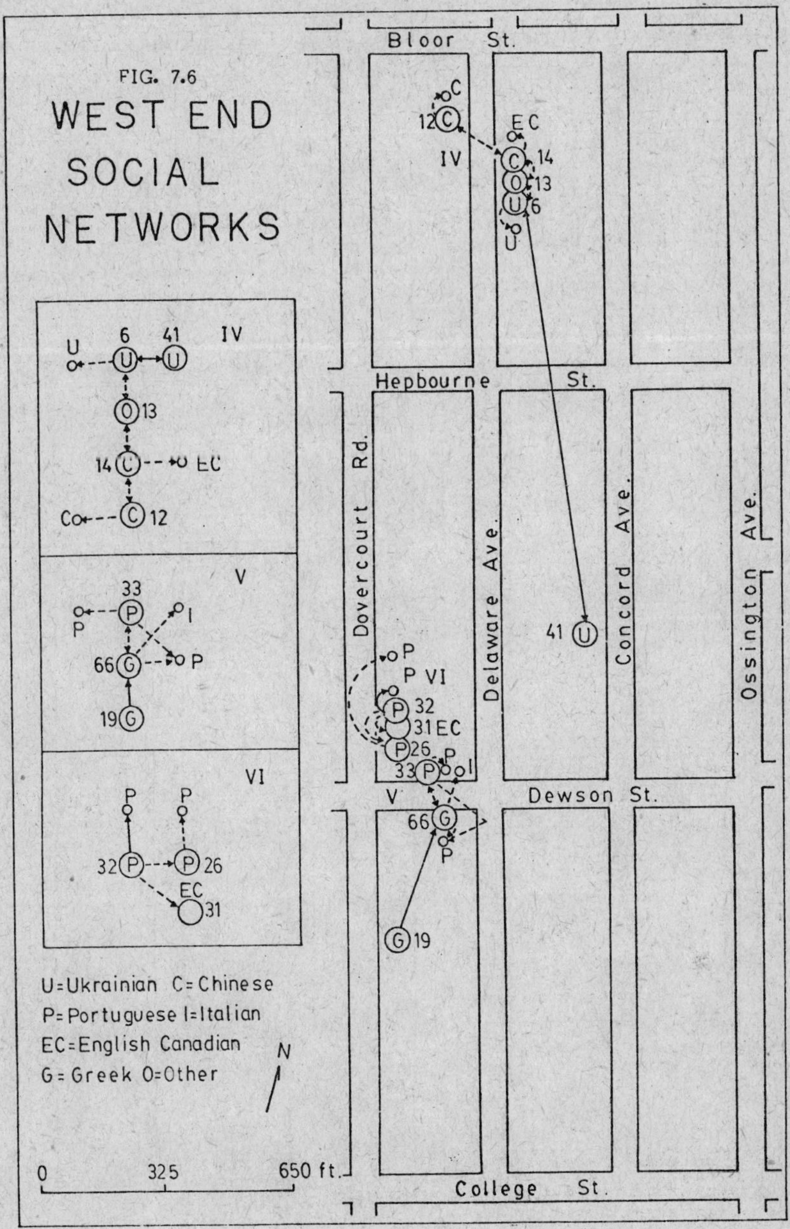

FIG. 7.6
WEST END SOCIAL NETWORKS

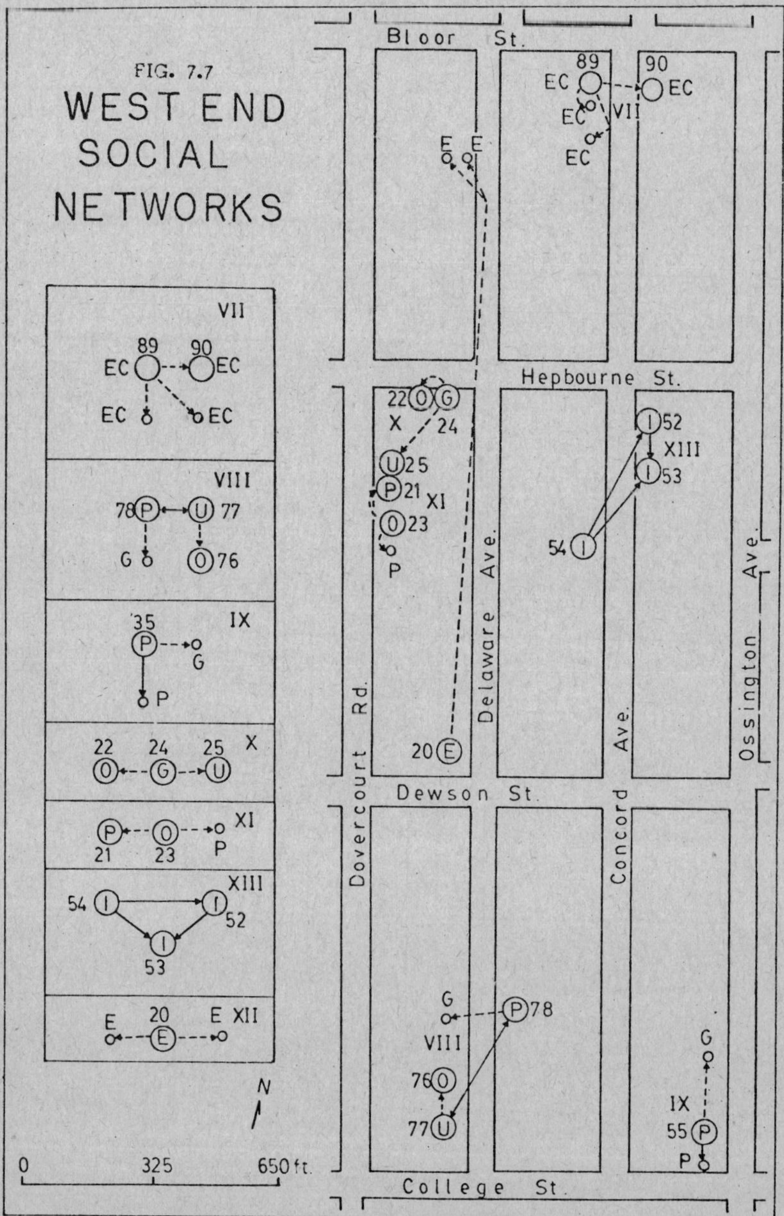
FIG. 7.7
WEST END SOCIAL NETWORKS
VII
89
90
EC
VIII
78
77
G
76
IX
35
P
X
22
24
25
XI
21
23
XIII
54
52
53
E
20
E XII
N
0
325
650 ft.
Bloor St.
Hepbourne St.
Dewson St.
College St.
Dovercourt Rd.
Delaware Ave.
Concord Ave.
Ossington Ave.
55

FIG. 7.8

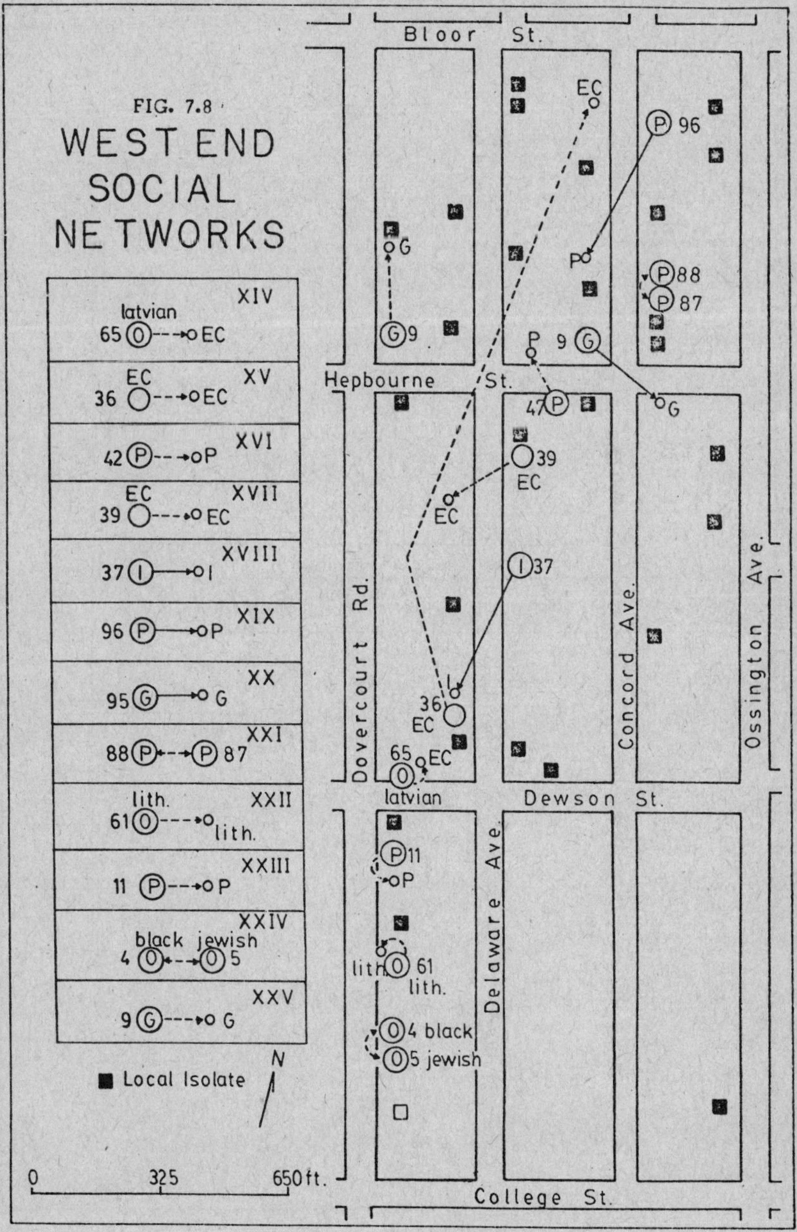
WEST END SOCIAL NETWORKS
XIV latvian 65 O ---→ EC
XV EC 36 ---→ EC
XVI 42 P ---→ P
XVII EC 39 ---→ EC
XVIII 37 I ——→ I
XIX 96 P ——→ P
XX 95 G ——→ G
XXI 88 P ←--→ P 87
XXII lith. 61 O ---→ lith.
XXIII 11 P ---→ P
XXIV black jewish 4 O ←--→ O 5
XXV 9 G ---→ G
Local Isolate
N
0 325 650 ft.
Bloor St.
Hepbourne St.
Dewson St.
College St.
Dovercourt Rd.
Delaware Ave.
Concord Ave.
Ossington Ave.

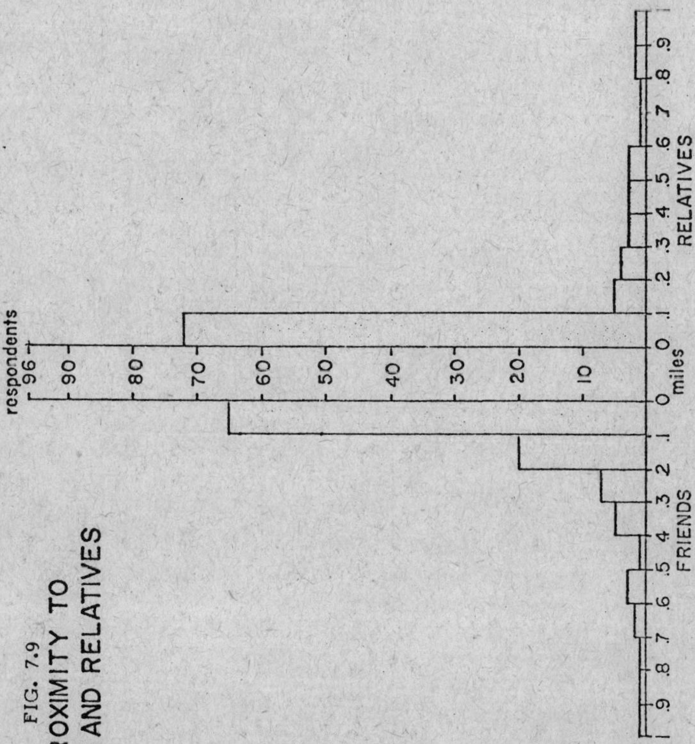

FIG. 7.9
PROXIMITY TO
FRIENDS AND RELATIVES

TABLE 7.3
Composition of Social Networks

Kind of Relationship

Friends	107	(70%)
Kin	46	(30%)
	N = 153	

Years Known	X	S_X
1st intimate	8	12
2nd intimate	4.2	6.2
3rd intimate	4	7.5

Religion

Catholic	94	(68%)
Orthodox	22	(16%)
Protestant	7	(5%)
Other	9	(6%)
None	7	(5%)
	N = 139	

Ethnicities

Portuguese	44	(33%)
Italian	40	(30%)
Ukrainian	4	(3%)
Greek	12	(9%)
Chinese	2	(1%)
East Indian	10	(7%)
English Can.	13	(10%)
Other	9	(7%)
	N = 134	

Occupations

Student	19	(11%)
Labourer	21	(16%)
Skilled Lab	25	(19%)
Semi-Skilled	12	(9%)
Prof/Manag	6	(5%)
Self-Emp.	16	(12%)
Retired	16	(12%
Unemployed	17	(13%
	N = 132	

Source: Field Interviews, 1977

It should not be forgotten that the basic control on intera ctions was social class. Most interactions were among peopl with similar class positions, although ethno-religious considera tions were the prime control on continuing social interaction.

Excluding diadic networks and dealing only with the thirtee networks of from three to fourteen members, several character istics are notable. A bit over half of the networks wer homogeneous in ethno-religious terms. Two networks were blu collar Italian Catholic. Two more were blue collar Portugues

Catholic. One network was middle class Greek Orthodox. One other was Ukrainian Catholic and one was mixed-ethnic middle class. Thus among those networks with high degrees of homogeneity, both class and ethno-religious composition were nonrandom. (Table 7.4)

TABLE 7.4
Network Summaries

A. Network Composition

#	(N)	Ethnicity	Religion	Relationships		Class
1.	(14)	50% Italian	62% Cath.	4 kin	8 fr.	2/3 blue collar
2.	(13)	60% Italian	70% Cath.	5 kin	9 fr.	2/3 blue collar
3.	(10)	30% Port	57% Cath. 40% Prot.	5 kin	6 fr.	60% blue collar
4.	(8)	40% Ukr.	40% Orth.	1 kin	6 fr.	60% blue collar
5.	(6)	Greek 75% Prot.	100% Orth. 35% Cath.	1 kin	4 fr.	middle class
6.	(5)	25% Eng. Can.	50% Prot.	1 kin	3 fr.	75% blue collar
7.	(4)	50% Eng. Can.			3 fr.	50% blue collar
8.	(4)	33% Ukr. 33% Ukr.	66% Cath.	1 kin	2 fr.	50% middle class
9.	(3)	Portuguese	Cath.	2 kin		blue collar
10.	(3)	2.3 Ukr,	Cath.		2 fr.	
11.	(3)	50% Port.	Cath.		2 fr.	blue collar
12.	(3)				2 fr.	middle class
13.	(3)	Italian	Cath.	3 kin		2/3 blue collar
	Diads after to #25 then isolates					

B. Network Groupings

#	Homogeneous (7)	Heterogeneous (2)	Intermediate (4)
2.	It. Cath.-blue collar	3. blue collar, local, kin core	1. It-Cath-blue collar (close to homogeneous)
5.	Greek, Orth, Middle Cl.	7. Eng. Can. blue/ middle, local	4. Ukr. (40%) blue collar
6.	Port-Cath, blue collar		8. Cath, 1/2 Ukr
9.	Port-Cath, blue collar		11. Port-Cath, blue collar
10.	Ukr-Cath.		
11.	Middle Class local		
13.	Italian-Catholic		

Source: Field Interviews, 1977

About half of the networks were heterogeneous in their social composition. Yet, even in these Italian Catholics, Ukrainian Orthodox, and English Canadians each comprised the core membership of a network. None of the heterogeneous networks could be considered random social aggregations. Where ethnicity or class were not dominant principals, kinship was.

However, the heterogeneous networks provided mechanisms for social integration in the local space. The largest network examined contained a stong core of Italian Catholics, but a quarter of its interviewed members were Orthodox, Ukrainian and Greek. Further, there was a Protestant in the network. The third largest network had a Portuguese Catholic core but included English Canadians and Greeks. Other networks included varying combinations of small ethnic groups with a core ethnic group. One combination that was rarely found included Italians and Portuguese. These were the dominant ethnic groups in the neighbourhood and appear to represent the basic ethnic cleavage in the area.

Kornblum has argued that heterogeneous social networks function to integrate varying ethnic groups in south Chicago.[18] I believe exactly the same is the case in West End, Toronto. About half of the respondents were embedded in homogeneous ethnic or class based social networks. The other half were parts of heterogeneous networks which linked categorical and, probably, sub-cultural groups around the basic ethnic cleavage in the neighbourhood. In fact, it appears that elements of ordered segmentation may be at the base of interactions.

When we turn to the spatial character of the networks, another dimension of control on structure emerges. Most networks were made up of very near neighbours. Typically, a given network was a blockfront group, often next door neighbours. When networks extended beyond a single block, most commonly the link was one of kinship. A common geographic pattern was a group of friendly near neighbours with kin based outliers on other blocks. Distances of interaction were short, and were based on diffuse neighbouring. Neighbouring, as a tendency to interact with others on the basis of residential proximity, was the foundation of social networks in the local territory.

In most networks both friendly and kinship bonds were found. Often, the topological core of networks was made up of a kinship core, with quite tight interlinkage, and a friendship periphery with much lower interlinkage. Geographically, however, the reverse was the situation. Most often friends were the geographic core of near associates and kin were located at some distance. In a number of other cases, however, there was quite close clustering of kin as well as friends.

Social composition, interactional structure and geographic

character all operated to produce a recognizable urban pattern of local interaction. That pattern involved the institutionalization of diffuse neighbouring in a complex web of visiting networks. Those networks tended to have a stable kinship and friendship base and a clear cut geographic form centered in the block-front group.

The census tract was not a defended neighbourhood. The study area was not dominated by either severe external problems or internal conflicts. Thus, neighbouring evolved into social networks with minimal neighbourhood structure. The geographic structure of networks does not aggregate into a territorially bounded neighbourhood. Instead, the study area appears as a very loosely knit congery of sub-neighbourhood segments.

Larger neighbourhood institutions did exist, but at a very low level of mobilization. Only fifteen per cent of the respondents belonged to voluntary associations in the local area. Those respondents belonged to local school and religious associations. Although the study area was not mobilized, several bases for potential mobilization were found.

Many more respondents belonged to voluntary organizations beyond the study area. These, however, were much of the same sort as those within the neighbourhood. Geographically, the outside voluntary associations tended to be bounded by the West End working class district of Toronto.

It would appear, on the basis of voluntary association memberships, that West End was the macro-neighbourhood or community of reference for the interviewed population. Within and outside of the local space, associational participation cut across the population of local isolates. In both local and non-local space local social isolates had a slightly higher level of associational participation than those embedded in local social networks. Thus, network isolates were not necessarily isolated even in local space.

Urban Sub-Neighbourhoods

The pattern of neighbouring found in the West End study area is a typical urban pattern. Its constituents are a geographical block front unit including both friends and relatives. Interactional distances and frequencies are not very great. However, the interlinkage between network members is quite high. At least the core of the block front network may be considered a clique.

This pattern of social interaction is found in many urban situations. Whyte found it in Levittown.[19] Recent studies have found it in middle class cottaging communities as well as in middle class Toronto and in the Toronto exurbs.[20]

In the West End study area ethno-religious factors were a primary local control second only to class in network structure. The ethno-religious control worked in a complex manner to integrate and divide the local area. About half of the sampled population belonged to homogeneous ethno-religious social networks. The other half of the interviewed population belonged to heterogeneous social networks. These heterogeneous networks commonly had an ethno-religious core, but integrated quite disparate groups. However, the two dominant ethno-religious groups, Italian Catholics and Portuguese Catholics, were together in a social network in only two cases.

I interpret this network evidenced cleavage between Italians and Portuguese as an indication of the existence of ordered segmentation among the West End respondents. This interpretation is advanced cautiously because it creates a number of theoretical problems. Ordered segmentation is the principal of social organization usually associated with mechanical solidarity, rural society, urban villagers or slum dwellers.[21] Ordered segmentation is usually part of a mobilized and defended neighbourhood in the context of a larger mobilized community of limited liability. In the city, it is associated with rapid invasion succession processes, and often with slum conditions which force the inhabitants to construct a locally based social order.

These are not the conditions which characterize the West End. Slow invasion-succession processes are underway, Italians and Portuguese are the primary movers into the area. They appear to be in competition for the recognized possession of the area. Both groups seek to have the territory recognized as "theirs", without attempting to displace the other. Ordered segmentation involves a series of oppositions and alliances between territorially organized groups at increasing scales of interaction. In this study only the lowest level has been examined. Larger patterns of oppositions, not measured in this study, exist.

For a significant portion of the sample, voluntary associations, both in the study area and beyond, provide a supplement, or an alternative, to visiting social networks. The combination of social networks, with their segmental cleavage, and voluntary associations provides the basis for future neighbourhood mobilization should the need arise.

Thus, some residuum of mechanical solidarity, local ethnic segmentation, continues amidst the organic solidarity based on the industrial-urban division of labour.

Ethnicity and religion, those sticky primordial bases of social life, continue. Sub-culture continues to differentiate populations even where the more modern class divisions based on income, education, occupation and wealth are operative.

Conditions in the West End of Toronto are relatively peaceful. The working class districts are not threatened with destruction under the guise of urban renewal or development. Further, internal social order is not greatly threatened by intense poverty and its associated evils. West End working people are not compelled to construct a local social order to survive. The larger urban infrastructure provides a great deal of continuing support.

Localism is, for the moment, relegated to a secondary position in social life. However, a lively world of social interactions continues to exist and function as a part of the social support and territorial system. Ethnic diversity provides opportunities for continuing, and even creating, local bonds based upon mechanical solidarity in a world interpenetrated by organic solidarity.

Should the larger situation alter, should the wreckers ball be put back to work, should safety become problemmatic, West Enders have the social resources to mobilize and form defended neighbourhoods and communities of limited liability. At this time the need is not strong and the block front networks suffice for individual and household support.

Acknowledgements

The data for this study was collected by honours students in Geography and Urban Studies during the winter of 1977. Especially important was the conceptual and analytic work of Beverly Verney. Computational work was overseen by Charles Matthews of the Geography Department of York University.

FOOTNOTES

1. J. Boissevain, **Friends of Friends** (Oxford, Basil Blackwells, 1974).

2. Works central to this tradition are: S. Leinbardt, editor, **Social Networks** (New York, Academic Press, 1977); J.

Boissevain and J.C. Mitchell, editors, **Network Analysis** (The Hague, Mouton, 1973); E.O. Laumann, **The Bonds of Pluralism** (New York, Wiley Interscience, 1973).

3. Suzanne Keller, The **Urban Neighborhood** (New York, Random House, 1968).

4. Gerald Suttles, **The Social Order of the Slum** (Chicago, University of Chicago Press, 1968).

5. For a perceptive review of the literature, T.F. Saarinen, **Environmental** Planning (Boston, Houghton Mifflin, 1976), Chapter 4, pp. 69-100.

6. Barry Wellman, **Urban Connections**, Research Paper No. 84, Centre for Urban and Community Studies and Department of Sociology, University of Toronto, 1976.

7. Gerald Walker, "Social Networks and Territory in A Commuter Village," **Canadian Geographer**, xxi, 4, 1977, pp. 329-350.

8. **Perspectives** Canada II (Ottawa: Statistics Canada, 1977), pp. 189-238.

9. John Swell, Up **Against City Hall** (Toronto: James Lewis and Samuel, 1972).

10. Walker, op. cit.

11. The basic ideas stem from the work of Emile Durkheim, **The Division of Labour in Society** (New York, Macmillan, 1933) editor, but have recently been reformulated by Gerald Suttles, **The Social Order of Community** (Chicago, University of Chicago, especially the idea of the defended neighourhood (1972).

12. Suttles, ibid.

13. William Kornblum, **Blue Collar Community** (Chicago, University of Chicago Press, 1974).

14. Laumann, op. cit.

15. Statistics Canada, Toronto: **Population and Housing Characteristics** (Ottawa, 1973).

16. Statistics Canada, op. cit., Part B, p. 55.

17. Walker, op. cit.

18. Kornblum, op. cit.

19. W. Whyte, **The Organization Man** (New York, Simon and Schuster, 1956).

20. Personal communications on three York University theses, **Social Network within a Suburban Setting** (York University, Department of Geography, 1974), B.A. Thesis, Richard Deyholos, **Social Interaction in a Cottage Community** (York University, Department of Geography, 1975), M.A. Thesis, Paul Cerre, **Behaviour and Social Interaction Patterns Among Cottagers** (York University, Department of Geography, 1974), B.S. Thesis.

21. Suttles, ibid.

EIGHT

Pierre deVise

Forces and Patterns of Segregation and Areal Differentiation in American Cities

The growth of urban areas in the Western World has been subject to various economic, social, and political forces which have resulted in very distinct patterns of segregation or areal differentiation. These forces involve spatial processes and result in spatial patterns which urban students, particularly sociologists, geographers and economists, have incorporated into urban models and theories.

Let us first define the population or characteristic which is subject to segregation or areal differentiation. The term "population" refers to any concentration of human beings identified by commonality of age, sex, race, ethnicity, family status, income, or labour force status; it also refers to businesses, institutions, individual animals, and plants. This sub-population can be assigned a given location at a given point in time. It has a spatial distribution. Such sub-population forms "natural" and "social" areas. A combination of such areas forms a spatial structure. Different areas are linked by channels of association and communications. There is spatial interaction. Sub-populations are spatially and functionally related to each other and to the total population made up of all sub-populations contained in a series of increasing hierarchic areas ranging from the neighbourhood to the entire world. Finally, there are forces, processes and patterns that explain areal differentiation and structure.

People who study such phenomena are known by disciplinary names like population or urban geographer, urban sociologist, and location economist or are known by interdisciplinary names like human ecologist, or regional scientist.

Ratzel, Durkheim, and von Thunen are nineteenth century

pioneers of, respectively human geography, social morphology and location economics. There was a parallel but independent development in human ecology and urban geography in the first half of the 20th century that did not converge until mid-century. Both developments started with a concern for explanations of forces and processes of segregation, then retreated into description and measurement in the face of mounting criticism and revisionism of their theories and constructs, and presently are grouping for larger theoretical constructs that can encompass all three components of ecological concern: distribution, spatial structure, and explanation.

A brief historical review of human ecology must begin then with general explanations of ecological forces, proceed to spatial structure, then pass to description and measurement, and end with new attempts at explanations of these forces.

The model of plant ecology inspired early 20th century sociologists of the Chicago School (Park, 1925) and environmentalist geographers (Hettner, 1895).

In their classic treatise on The City, Park, Burgess and McKenzie outlined an ecological approach to the study of urban communities. Inspired by the model of plantecology, they saw urban growth as a progression from the "simple to the complex, from the general to the specialized; first to increasing centralization and later to decentralization." Community growth is accompanied by differentiation and segregation of land uses and occupants. Just as in plant communities, growth takes the form of repeated invasions of land uses and occupants; there is a new cycle of adjustment either generating new growth or retrenchment. Undesirable invaders include "foreign races" and individuals who "are free from the customary restraints of family life." Just as weeds make inroads into flower gardens, these human invaders enter at points of entry of highest mobility and lowest resistance. The former occupants are driven out during the initial, developmental and climatic stages of invasion. Land uses and institutions adjust to the needs of the new population. Each "natural area" or "moral region" develops attributes that attract congenial human groups and repel others.

Although we think of racial and ethnic colonies as the major elements of social differentiation, age and sex segregation also takes place in this filtering process. The authors found that, in the central core, males outnumber females by 3 to 1 to 5 to 1. But the sex ratio favours females in the outlying residential neighbourhoods. The "settler type of population" is driven outward by the "more mobile and less responsible adults" who "herd together in the hotel and apartment regions" of the urban core. The settler population serve as the "custodians of the stabilizing and repressive mores." They are the "conservative, law-abiding, civic-minded population." In contrast, the individuals who live in the core are "a class of people who are not only more mobile but whose mores and attributes are more vagrant and radical."

These eccentric and exceptional people, because of their rejection of the moral restraints and inhibitions of normal family life, find in other abnormal groups a stimulus and a moral support. Thus do the eccentric, the poor, the vicious and delinquent become particularly fit for the "unhealthful and contagious intimacy" in which they are "condemned to exist".

To Park, Burgess and McKenzie, the central research question was "howfar the increase in crime keeps pace with increasing mobility and to what extent this mobility is a function of the growth of population". Their tentative answer is that it is probably the processes of social disorganization and moral disintegration unleashed by individuals breaking away from their families which are responsible for the increase of vice and crime in large cities. The intermixture of the eccentric, the poor and the vicious is a "fascinating but dangerous experiment". Elements of chance and adventure add to the stimulus of city life and attract the "young and fresh". The lure of cities may be explained, "like the attraction of the flame to the moth, as a sort of tropism".

In a companion volume to The City entitled "The Urban Community" and edited by Burgess, Harvey Zorbaugh profiled the rooming-house area which, like other "natural areas," tends both to sort and stamp its population. The rooming house area is a childless are of young, motly single, adults who work in mostly white-collar jobs. It is an area of extreme mobility and anonymity which thwart ambition and "destroy public opinion." The single person is socially isolated; he finds neither "security, response nor recognition". He is "restless and he is lonely". His behaviour takes one of three forms: day-dreaming to satisfy repressed wishes, fetishism or the transfer of affections to symbols and objects, and suicide "frequently the only way out to the persons of the rooming-house world". In any case, old associations are ended; old standards disintegrate; behaviour is impulsive rather than social, and social control collapses.

Writing in 1925, at a time when census characteristics were collected by political subdivisions "which have no real correspondence with natural areas of the city," Zorbaugh stressed the need for the collection of social statistics at the level of "natural areas." Such statistics, he believed, "should prove diagnostic of real situations and processes, indicative of real trends". In later years, Zorbaugh participated with other members of the Local Community Research Community in defining what developed into the present subdivision of Chicago into 76 community areas and 822 census tracts.

In 1940, the Census provided for the first time census tract tabulations of social and mobility characteristics of the population of large cities that permit the testing of the hypotheses of Park, Burgess and McKenzie. Ronald Freedman, a student of Burgess's, undertook such a survey using Chicago as the study area.

Using mobility and rent data reported by the 1940 Census, Freedman divided the city in 24 zones and identified within these migrant/high rent, migrant/low rent, nonmigrant/high rent, and nonmigrant/low rent areas. He determined that migrant/high rent areas formed a solid tier along Chicago's lakefront, backed up by migrant/low rent areas. The high rent areas made up the primary areas, and the low rent areas, the secondary areas of migrant concentration.

In confirmation of his hypotheses, Freedman discovered that the migrant areas were characterized by central and transient housing extending out from the city centre along a rapid transit corridor, and that migrants tend to settle in areas of the migrant zone with socioeconomic characteristics of the nonmigrant residents similar to their own. Freedman next compared the spatial distribution of migrants with that of indicators of social disorganization, juvenile delinquency, mental illness and family disintegration. He found no confirmation for previous studies of the Chicago school which hypothesized correlation between mobility and these indicators.

Freedman concluded that most migrants to Chicago in the years 1935-40 occupied a "relatively favourable economic and social position". Especially singles, who made up a fourth of the migrants, and couples without children had socioeconomic characteristics and styles of life associated with the "ideal-typical urban person". Freedman speculated that the settlement pattern of recent migrants differed considerably from that of foreign, immigrants of previous decades. In the new settlement pattern, mobility is not disorganizing to family and community life. Moreover, new settlement patterns tend to support the wedge-sector concept of urban growth, rather than the concentric circle or gradient concept suggested by earlier studies of social disorganization.

Freedman is but one of many empirical and theoretical critics of classical ecological theory. Firey (1945) and Hatt (1946) pointed to theoretical weaknesses including the confusion between biotic and cultural elements, the emphasis on competition as the basis of social organization, and the lack of generality of concentric zonation and natural area. As a result, urban sociologists retreated into measurement and classification of urban phenomena.

Social area analysis was one of the better known techniques of grouping interurban subareas according to common characteristics. Shevky and Bell (1955) devised three indexes, made up of census tract variables, measuring social rank (economic status), urbanization (family status), and segregation (ethnic status).

The method of classification was that of orthogonal division into sixteen social areas made up of four by four equal parts of the urbanization and social rank indexes. Use of the segregation index was made optional. Although all tracts in Los Angeles and San Francisco were listed under the various social area designations, the social areas were not mapped, and there was no discussion of their spatial distribution.

What of the relation between social areas and behaviour? Bell argues that the social character of local areas within a city as defined by economic, family and ethnic characteristics is an important predictor of individual attitudes and behaviours, subcultural patterns and social organization. Another purpose of the social area classification was to provide a framework for sampling design.

In response to criticism on theoretical and empirical grounds, Bell (1955) used factor analysis to validate the three-construct classification of social areas and Van Arsdol (1958) found six more cities that confirmed the social area measures, and four cities that did not.

State Economic Areas (Bogue, 1951) and "homogeneous regions" (James and Jones, 1954) used somewhat similar techniques and concepts to those of social area analysis. Using counties as building blocks, Bogue delimited economic areas in terms of grouping contiguous counties displaying multi-factor homogeneity.

So far, social areas and economic areas show little if any relation to urban and regional models of spatial structure specified by the earlier generation of urban sociologists, geographers and economists. Homogeneous regions were defined as single-feature or multiple feature regions in a "regional pattern that has both meaning and significance when it can be interpreted in terms of systematically related processes, operating through time". (Brave words for a path that turned into a dead-end.)

Rees (1970) applied factor analysis to intra-urban data in a technique called "factorial ecology". For his study of the Chicago SMSA in 1960, Rees constructed a data matrix containing 57 census tract variables for each of 222 subareas, compressing the common elements of variability into seven dimensions that additively reproduce this common variance. The principal dimensions of variation in socioeconomic characteristics of the metropolitan population are social rank, family status, racial status, and foreign stock status. These dimensions vary systematically across the physical space of the metropolis. Social rank varies both with direction and with distance from the city centre. Social rank also varies systematically with distance from the Loop. Family status also exhibits zonal and sectoral variation. There is the same rise in family status away from the city centre with a peak again in the outer suburbs and a fall-away as the rural margins of the urban regions are reached. Racial status shows extreme clustering, with an extremely high degree of mutual segregation between racial groups. Most of the black population of the metropolitan area (91 per cent) live in the city of Chicago. Foreign stock status exhibits within the city a distribution of black neighbourhoods. Application of grouping techniques to the distribution of the communities in a space defined by these principal dimensions leads to a classification of these communities in terms of community-of-interest types.

In figure 8.1 taken from the Rees study, the basic triad of spatially arranged social dimensions can be superimposed to form, at the intersections of sectors (A), zones (B) and black ghettoes (C), an additive composite spatial model of the Chicago metropolis (D). However, the zones within the ghettoes do not correspond to the zones of the larger metropolis. The ghettoes reproduce in microcosm the metropolitan-areawide pattern (E). This modified model is further distorted by the way the city grew (F), producing tear faults in the zones (G). Finally, the introduction of secondary work-place nodes such as the industrial suburbs and satellites, and the heavy industry nucleus (H), and their coalescence over time with the major node present a fifth dimension to be integrated in the composite model (I).

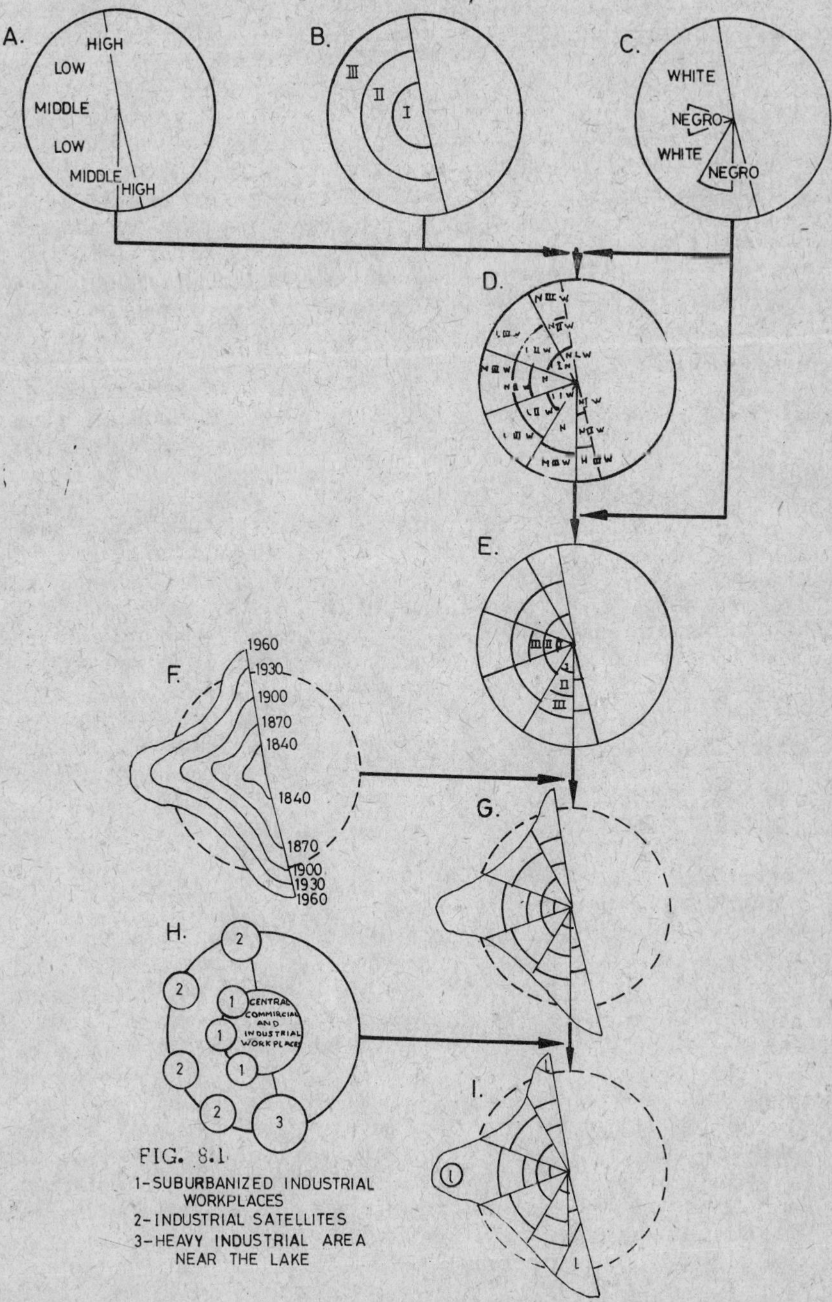

FIG. 8.1

1-SUBURBANIZED INDUSTRIAL WORKPLACES
2-INDUSTRIAL SATELLITES
3-HEAVY INDUSTRIAL AREA NEAR THE LAKE

In his Chicago study, Rees confirmed the importance and independence of the three dimensions identified in social areal analysis. He also sought to reconcile zonal and sectoral models of urban structure which dominated in classical ecology.

We recall that, according to Burgess, McKenzie and Park, the city grows as social groups move outwards as they move up the social ladder, creating a patterning of the social groups of the city along zonal lines. Hoyt (1939) postulated an alternative view that the key mechanism structuring the ecology of the city was the behaviour of the high rent portion of the real estate market, which expanded outwards in a sectoral manner along transport or amenity axes. Resolution of this seeming conflict was accomplished by the recognition that there was not just one mode of spatial differentiation of the socioeconomic characteristics of metropolitan populations but several, and that underlying these spatial patterns was a process of individual choice linking the distribution of housing types and costs with ethnic types.

In Chicago and other large American cities, the three-factor social area construct of status, familism and race is adequate to differentiate areas. The more complex ethnicity of non-Western cities requires modification of the American model. In the Indian case, differences between regional origin, between Hindu and Moslem, and between castes require an addition of several geoethnic factors. Berry (1977) and others used factorial ecology to reveal four types of social areas in Calcutta made up of cross-ratings of Factor 1, a land-use and familism gradient, and Factor 4, a social status gradient inversely related to scheduled caste.

In figure 8.2, we can compare the distribution of social areas in the social status-family status space in Chicago and Calcutta. Chicago's high status social areas are positioned in the mature lakefront sector (A) and in the young suburban ring (B), whereas Calcutta's old and young high status areas (A & B) are clustered in the commerical core. In contrast to Calcutta, in which status and caste are inversely related, Chicago's ghettoes, we repeat, form a distinct and independent sector (D).

Because of more complex links between region of origin, religion, caste and occupation, several land-use/ethnic factors are needed to differentiate status in Calcutta--Factor 3, literacy; Factor 4, high-status residential areas; Factor 5, middle-status non-Bengali commercial areas; and Factor 6, middle-status Bengali commercial areas.
seeming conflict was accomplished by the recognition that there

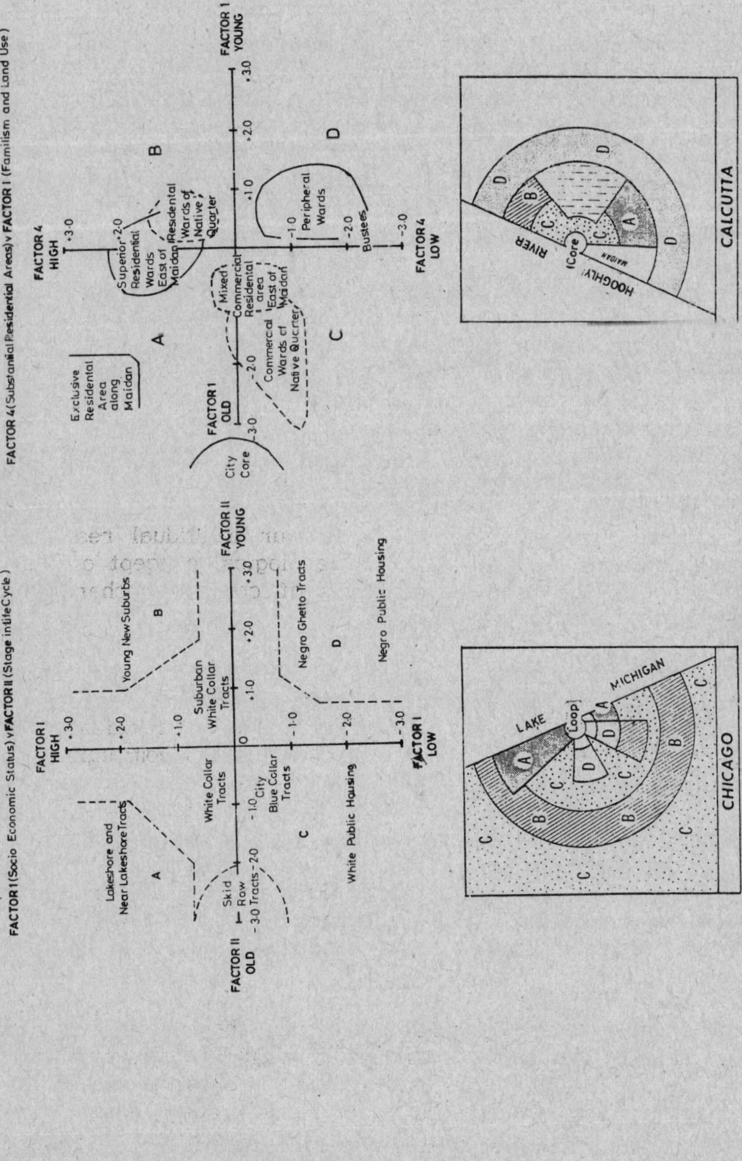

FIG. 8.2 GENERALIZED SOCIAL AREAS OF CHICAGO AND CALCUTTA

After Rees via Berry, Comtemporary Urban Ecology (1977)

was not just one mode of spatial differentiation of the socioeconomic characteristics of metropolitan populations but several, and that underlying these spatial patterns was a process of individual choice linking the distribution of housing types and costs with types of individuals and families who lived in them.

In common with the critique addressed to social area analysis, the technique of factorial ecology was criticized on both theoretical and empirical grounds. Alford (1972) acknowledged that factor analysis was useful as a device for reducing a complex and large number of phenomena across many su.areal units to a small number of independent dimensions. He questioned, however, the arbitrary selection of cities and variables, suggesting that availability of data and the interest of researchers dictated both choices. He also questioned the presumption that the causes and consequences of the identified principal factors are indeed independent of each other. Concern for casual processes should have dictated the selection of variables, the sample of units, and the interpretation of the data, he said. Factorial ecology, Alford concluded "cannot serve as both the beginning and the end of analysis, starting with unquestioned arrays of input characteristics and ending with unexamined labels for the output factors".

Factorial ecologists tended to favour individual residential location decisions over the classical ecological concept of group competition as the behavioural basis of community characteristics.

The concept of "social space" as elaborated by Chombart deLauw e (1952) was shown by Berry (1977) to illustrate to role of household residential location in the determination of social status, family status and ethnic status of communities, as per the following scenario:

Households occupy a location in social space depending on economic status and family status. The social space includes housing space, and community space. From a range of acceptable communities, a location is chosen in proximity to the work place, and other desired amenities, thus fixing the choices in physical space.

An orderly social ecology evolves by the selection of like spaces by like households, resulting in segregation by economic status and by family status. Blacks make up a microcosm of the metropolitan area because they are excluded from white communities both through racial discrimination and economic segregation.

A major overhaul of human ecology was attempted by Hawley (1950). How humans organize and adapt to a constantly changing environment is the major question to be answered by ecologists, Hawley said. He identified three major dimensions of social organization:

(1) Interdependence along two axes: symbiotic and commensalistic. Division of labour is an example of symbiotic interdependence and social groups with a common interest exemplify commensalistic interdependence.
(2) Organizations transcending individual actors and characteristics.
(3) Tendency to functional, demographic and distributive equilibrium.

Stability and change may be studied though the "ecological complex" of four reference variables: population, organization, environment and technology, according to Duncan (1959). Because these four variables are functionally interdependent, a change in one will affect the other three. This interdependence lays the groundwork for interpreting change as part of a feedback mechanism within the ecological complex.

Matching the ecological revisionism of Hawley, urban geography underwent radical transformation in the 1960s with the application of location theory and the quantitative approach to the construct of the nodal region.

The central ideas of the nodal region are:

(1) Human occupance refers to the spatial organization of households and firms.
(2) Spatial organization centres around central places which, together with their functions and hinterlands, form nodal regions.
(3) These nodal regions are arranged in a hierarchy, determined by the number and type of functions provided by their central places.
(4) The higher ranking central places perform specialized functions which lesser places do not perform. These more specialized functions have a higher "threshold" - require a greater demand - then can be provided by smaller places. The size of the "threshold" determines the "range" over which these functions can extend their markets.
(5) The functional hierarchy is reflected not only in number

of functions, but also in the spacing of settlements, population densities, and land values. The settlements with fewer functions are not only smaller, they are also closer together than are larger settlements with more functions.

(6) The urban hierarchy is made up of alternating systems of nodal and uniformal areas. The central city and the satellite cities and villages are nodal. The suburbs fringing the central city tend to be uniformal: dormitory, industrial, resort or institutional suburbs. These form a "nested hierarchy".

(7) The spatial system of nodal regions forms an interlocking network of connections between larger and smaller regions distributed over the larger metropolitan area in a relatively orderly way depending on the intensity of functional localization. The boundaries between nodal regions and sub-regions are not precise, but they can be defined in some arbitrary fashion.

(8) Spatial interaction occurs among the nodal regions of all sizes and places in the hierarchy. Measurement and analysis of this interaction can take the form of origin-destination surveys and gravity models.

(9) A nodal region changes through time. Regional process involves both natural evolution and technological change. Regions change their size, form, and functions through natural growth and technological adaptation.

The nodal region is an abstraction. It can be used as model against which reality can be compared. In terms of economic analysis, the region is regarded as a trading and producing area possessed of a high degree of functional integrity, a measureable demand for raw materials and finished goods, and measureable economic relations with other regions. The region's exports and imports, its balance of payments, and its gross product can be measured.

In demographic terms, the region can be studied through an analysis of subarea socioeconomic residential characteristics, place of residence to place of work linkages, population growth linkages with employment growth, population density, linkages with transportation. In terms of land use analysis, the region can be examined through its morphological patterns of land utilization, land values, population densities, and settlement.

Paralleling ecologists' concern with social system change and maintenance, contemporary urban geographers see spatial structure operating on spatial behaviour through screens of cognition (Berry, 1977). According to Berry, spatial behaviour in adaptation to the environment "contributes to space-contingent processes that maintain a given steady state, to space-forming processes producing evolutionary change, or to space-transforming (revolutionary) processes that, to complete the circle, radically transform the environment".

REFERENCES

Alford, Robert R. "Critical Evaluation of the Principles of City Classification," Chapter 11 in Brian J.L. Berry (ed.) **City Classification Handbook**: Methods and Applications. New York: Wiley-Interscience (1972).

Bell, Wendell, "Economic, Family and Ethnic Status: An Empirical Test," **American Sociological Review**, 22 (1955) pp. 45-52.

Berry, Brain J.L. and John D. Kasarda. **Contemporary Urban Ecology.** New York: Macmillan (1977).

Boque, Donald J. **State Economic Areas.** Washington: USGPO (1951).

Chombart deLauwe, Paul H. **"Paris et l' Agglomeration Parisienne.** Paris: Presses Universitaires de France (1952).

Duncan, Otis Dudley and Leo F. Schnore. "Cultural, Behavioural, and Ecological Perspectives in the Study of Social Organization," **The American Journal of Sociology**, 65 (1959) pp. 132-146.

Firey, Walter. "Sentiment and Symbolism as Ecological Variables," **American Sociological Review**, 10, 2 (1945) pp. 140-148.

Freedman, Ronald. **Recent Migration in Chicago**, Chicago: The University of Chicago Press (1950).

Hatt, Paul. "The Concept of Natural Area," **The American Sociological Review** 11, 4 (1946), pp. 423-427.

Hettner, Alfred. "Die Lage der menslicken Ansidelungen, "**Geographische Zeitschrift**, 1 (1895) pp. 361-375.

Hawley, Amos H. **Human Ecology: A Theory of Community Structure.** New York: Ronald Press (1950).

Hoyt, Homer. **The Structure and Growth of Residential Neighborhoods in American Cities.** Washington: USGPO (1939).

James, P.E. and C.F. Jones (eds.) **American Geography: Inventory and Prospect.** Syracuse: Syracuse University Press (1954).

Park, Robert E., Ernest W. Burgess and R.D. McKenzie. **The City.** Chicago: The University of Chicago Press (1925).

Rees, Philip H. "The Factorial Ecology of Metropolitan Chicago," in Brian J.L. Berry and Frank H. Horton (eds.) **Geographic Perspectives on Urban Systems.** Englewood Cliffs: Prentice-Hall (1970).

Van Arsdol, Maurice, Santo F. Camillori, and Calvin F. Schmid. "The Generality of Urban Social Area Indices." **American**

Sociological Review. 23 (1958) pp. 277-284.

Zorbaugh, Harvey W. "The Dweller in Furnished Rooms: An Urban Type," in Ernest W. Burgess (editor) **The Urban Community,** Chicago: The University of Chicago Press (1926).

NINE

Christina R. Victor

Ethnic Minorities in Nottingham C.B.

A Study of Levels of Residential Segregation 1961-1971

INTRODUCTION

Studies of ethnic residential patterns have been reported for a variety of urban contexts. However, the majority of this work has been concerned with the cities of North America. Within a specifically British situation far fewer studies are in existence. This is especially true for investigations which have sought to review the residential segregation of immigrant groups at more than a single point in time. Of the major British studies, London has been the focus of several investigations (Lee, 1973, Dalton and Seaman, 1973) whilst Birmingham formed the study area for the work of Jones (1973; 1976). Thus there is a marked lack of published research focussing upon medium sized urban areas and their immigrant communities. In this particular investigation the nature and magnitude of ethnic minority group segregation is described for Nottingham, the twelth most populous town in Britain in 1971.

Peach (1975) has identified four phases in the spatial study of minority group residence patterns. There are descriptive analyses at macro and intra-urban scales, the dissimilarist approach and the application of predictive models to the study of racial groups. This particular work forms part of the dissimilarist tradition of research. In this statistical indices are used to measure the degree of segregation displayed by a particular ethnic group. For this specific investigation of Nottingham, segregation patterns are examined for a broad range of immigrant groups rather than a concentration upon a specific racial minority. Thus a comprehensive study of minority group segregation was undertaken, a task which has lost favour recently. Before the methodology and results are described, the following section summarises very briefly the nature of the theories which have been formulated to explain the residential segregation of minority groups.

Ethnic Segregation in British Cities

The Theoretical Background

The influences which combine to produce the segregation of ethnic minorities in British cities can, according to Jones (1979), be grouped into two distinct categories. The first group of influences, internal (or subculural) factors relate to the social aspirations and needs of the minority group. In contrast external (or cultural) influences identify the relationship existing between the aggregate ethnic group and the host society. Attempting to assess the balance between these two sets of factors is difficult, especially when the production of spatial patterns of segregation is being considered. However, by briefly describing the nature of each influence, the relative characteristics and strengths of the two models are identified.

Subcultural or positive segregating influences stem from the cultural distinctiveness of the ethnic group. Ethnic minorities are usually recently arrived immigrants who differ markedly, in cultural terms, both from each other and the host population. It is suggested, therefore, that voluntary residential propinquity, brought about by a desire to preserve both formal and informal cultural institutions, leads to geographical clustering and segregation from the host population.

The relationship between the ethnic group and the institutions and structures of the host society identify the negative forces of residential segregation. Although formal discriminatory barriers on grounds of race, are illegal, there can be little doubt that ethnic minorities have been handicapped in the competitive markets which make up contemporary British society. Discrimination is at its most evident in the fields of employment and housing. In terms of housing, British ethnic minorities have been hindered by the operation of two discreet sets of factors Firstly, their low incomes have restricted their ability to 'compete' for housing in the private sector. In consequence these groups have been consistently identified with substandard housing. A second set of factors helping to maintain the residential separateness of ethnic minorities are the difficulties surrounding the entry of such groups into the public housing sector. Opportunities for creating more ethnically heterogeneous residential distributions, via the use of the public housing stock, have largely been ignored.

Data Sources

The data sources used in this study consisted of the birth place tables derived from the 1961 and 1971 census enumerations. Whilst Cater and Jones (1979) have criticised the use of census data for studying minority group segregation, their preferred data source, the electoral roll also has inherent limitations. By employing the electoral register, any research is restricted to groups such as Indians and Pakistanis who can be readily identified by their distinctive names. Studies such as this one, which seek to review residential patterns for all major immigrant groups are thus constrained to use official census returns.

It should, however, not be assumed that the deceennial census returns are without limitations. Fundamentally the census records information relating to country of birth of residents. Information relating to racial or ethnic origins in the strict sense of the terms is not collected. Hence the researcher is obliged to make a number of assumptions regarding the nature of the information. For example, whilst it is assumed that all those born in Africa are "non-white", there are obviously cases where this is incorrect (Husain, 1975). In addition, country of birth is not always a correct guide to the ethnic status of individuals.

A further limitation is that the information related to residents of England and Wales. The definition of residence employed by the census remains, however, somewhat subjective. An overseas student, or worker living in Britain, could state that his residence was his home country. This has resulted in the consistent under-enumeration of the immigrant community in Britain (Peach and Winchester, 1974).

For this particular investigation, the individual countries of birth of residents were aggregated into a number of major groups. The Irish groups unites those born in Eire and Ulster whilst the European born group identifies those born in mainland Europe. Included within the old Commonwealth group are those residents born in Canada, Australia and New Zealand. The final group identified relates to those born in the New Commonwealth, i.e., India, Pakistan, the West Indies and Africa. These groups, rather than the more extensive and analytically more complicated individual countries of birth, form the basis of the analysis described later.

Spatial Framework

A persistent problem in the study of minority group residential patterns is the geographical scale at which the investigation is undertaken. As described later, this problem also has ramifications in terms of analytical procedure. If the size of the areal units employed is too large, the realities of minority group segregation may well be submerged. Conversely too fine a mesh of spatial units may well overstate the extent of segregation.

When attempting to study changes in ethnic residential distributions over time a second problem is encountered. As Cater and Jones (1979) note correctly, changes in the boundaries of the spatial units used to collect census information has made it almost impossible to study minority group residence patterns in a dynamic manner. However, in this investigation, a framework of medium sized areas, which were directly comparable in both 1961 and 1971, was employed. For Nottingham C.B., both the sets of enumeration districts defied in 1961 and 1971 could be aggregated to form the 140 E.D's used for the 1966 sample census. By expressing the two data sets in terms of this constant areal framework it was then feasible to study continuity and change in the residential structure of Nottingham's ethnic community.

Analytical Procedure

The statistical measurements of residential segregation, via the use of the dissimilarity and segregation indices, is currently an area of considerable contention in geographical research. Peach (1979) and Jones (1979) have debated this problem in considerable depth. As Woods (1976) correctly states the central problem with these indices is again that of 'scale'. Given this limitation, which also applies generally to all spatial analysis, Jones (1979) concludes that these measures form the most appropriate and comparable methods for the measurement of residential segregation at the present time.

Following Duncan and Duncan (1957), the two measures of segregation used in this study were the index of dissimilarity (I.D.) and the index of segregation (I.S.). Described briefly the index of dissimilarity expresses the proportion of a population sub-group, Y, that would have to redistribute between a set of N sub-areas to illustrate a distribution equal to that of another sub-group, Y. In contrast, the index of segregation describes the proportion of a population sub-group, Z, which would have to redistribute between N sub-areas to produce a distribution equal to the total population, T.

For each of these two indices, segregation is measured on a continuous scale ranging from 0 (no segregation) to 100 (total segregation). Both these indices were calculated, for 1961 and 1971, for the major birth place groups identified previously. Prior to the presentation of these results, the growth of immigrant community, between 1961 and 1971, is briefly described.

The Growth of the Immigrant Community, 1961-1971

The total population of Nottingham C.B. decreased by 11,224 persons (3.6%) between 1961 and 1971. During this same period Table 9.1 indicates that the foreign born population increased by 6,394 persons (42.4%). This represents an increase in the relative size of the ethnic community from 4.9% (1961) of the total population to 7.1% (1971). This large increase predates the imposition of strict immigration quotas. As a result it is doubtful if the next enumeration in 1981 will identify such a large increase in the size of the foreign born community. It should be noted that no account is taken in these figures of the Britich born ethnic community sector of the population likely to increase in importance in the future.

During the decade 1961-1971, the structure of the ethnic community displayed some important changes. These are summarised in Table 9.2. In 1961, the foreign born community was neatly divided into three groups of similar size - the Irish, New Commonwealth and European born groups - with the Eire born group the largest ethnic minority.

By 1971, this structure had been modified. Over the decade the number of residents enumerated or born in either Ireland or Europe remained stable. Irish migrants remained principally those born in Eire. An interesting change in the country of birth of European migrants is, however, evident. Of new immigrants the numbers originating in Western Europe (i.e., Poland, Italy and Germany) declined markedly and were replaced by the 'other' Europe group (i.e., those born in Spain, Portugal, Greece, etc.). Overall, therefore, the European and Irish born immigrant group declined in importance in terms of the total foreign born community.

Predominantly, the decade 1961-1971 was characterised by a massive growth in the size of the New Commonwealth group. In 1961, this group represented 33.6% of the foreign born community. By 1971 this proportion had risen to 52.1% of the foreign born community with the British Caribbean Group constituting the largest single ethnic group. Substantial increases were also recorded in the number of residents enumerated as born in Africa, India and Pakistan. Clearly, therefore, the increase in the size of the foreign born community in Nottingham

TABLE 9.1
Nottingham C.B.: The Immigrant Population, 1961-1971

BIRTH PLACE	1961 TOTAL	1961 % POPULATION.	1971 TOTAL	1971 % POPULATION	CHANGE	INDEX OF CHANGE
IRELAND	5062	1.62	5403	1.80	+341	106.7
Ulster	1173	0.38	1413	0.47	+240	120.4
Eire	3889	1.25	3990	1.33	+101	102.5
OLD COMMONWEALTH	378	0.12	404	0.13	+ 26	100.8
Australia	120	0.04	130	0.04	+ 10	108.3
New Zealand	0	0.00	39	0.01	+ 39	390.0
Canada	258	0.08	235	0.08	- 23	91.0
NEW COMMONWEALTH	5065	1.62	11182	3.72	+6117	220.7
Africa	247	0.08	871	0.29	+ 624	352.6
India	996	0.32	2389	0.80	+1393	239.8
Pakistan	362	0.12	1970	0.66	+1608	541.1
British Caribbean	3305	1.06	5631	1.87	+2326	170.3
'Other' Asian	155	0.05	324	0.11	+ 169	209.0
EUROPE	4560	1.46	4470	1.49	- 90	98.0
Poland	2095	0.67	1725	0.57	- 370	82.3
Italy	880	0.28	600	0.20	- 280	68.1
Germany	807	0.26	860	0.29	+ 53	106.5
'Other' European	778	0.25	1285	0.43	+ 507	165.1
TOTAL	15065	4.83	21459	7.14	+6394	142.4

TABLE 9.2
The Structure of the Immigrant Community, 1961-1971

Birth Place	% total foreign born population 1961	1971
IRELAND	33.6	25.1
Ulster	7.1	6.5
Eire	25.8	18.5
OLD COMMONWEALTH	2.5	1.8
Australia	0.7	0.6
New Zealand	0	0
Canada	1.7	1.0
NEW COMMONWEALTH	33.6	52.1
Africa	1.6	4.0
India	6.6	11.1
Pakistan	2.4	9.1
British Caribbean	21.9	26.2
'Other' Asia	1.0	1.5
EUROPE	30.2	20.8
Poland	13.9	8.0
Italy	5.8	2.7
Germany	5.3	4.0
'Other' European	5.1	5.9

between 1961 and 1971 was mainly attributable to the growth of the New Commonwealth community.

The Spatial Segregation of Ethnic Communities, 1961-1971

That ethnic minorities are residentially segregated within British cities has been identified by several authors (Husain, 1975; Jones, 1979). However, much less attention has been given to variation in segregation levels between the various ethnic groups. There is a strong tendency, within the British literature, to ignore the important social and cultural differences manifested by the various minority groups. By examining the degree to which minority groups are segregated from each it is possible, at least superficially, to identify the extent to which positive culturally based factors of segregation have influenced the residential location of individual groups.

For this particular exercise, the individual countries of birth were aggregated to form four major cultural groups. These groups were the Irish, European, Asian (i.e., India and

Pakistan) and West Indian birth place groups. Each group was considered to have a readily identifiable and well developed cultural identity. As table 9.3 demonstrates, considerable differences in residential location were characteristic of the Asian, West Indian and European birth place groups in 1961. However, an important degree of residential similarity is experienced between the Irish and European. Clearly, therefore, this table does suggest that, in 1961, sub cultural influences were important positive influences upon the specific residential location of individual minorities.

TABLE 9.3
Nottingham C.B.: Index of Dissimilarity—1961

	Birth Place	1	2	3	4
1.	Ireland	-			
2.	Asia	37	-		
3.	West Indies	36	42	-	
4.	Europe	19	34	42	-

Inter-group dissimilarity measures were then calculated for 1971. It was hypothesised that segregation levels would have decreased over the decade. This conclusion was arrived at bacause of the extensive slum clearance policy which was activated during the period 1961-1971. Thus it was considered that, with a depleted private rented housing sector, ethnic groups competing for such housing would display a more similar residential distribution, regardless of cultural differences. However, as Table 9.4 illustrates, this hypothesis was incorrect. Inter-group segregation levels increased rather than decreased during this decade. This increased degree of residential dissimilarity was especially characteristic of the Asian Community. That segregation is not, however, a simple function of skin colour, is demonstrated by the increased dissimilarity measures recorded by the European born group. Clearly, therefore, at both points in time subcultural positive forces share an important influence upon the residential distribution of Nottingham's immigrant community.

Table 9.5 presents the Index of Segregation for these groups in both 1961 and 1971. This statistics provides a crude measure of overall segregation and may be interpreted as providing an approximate indication of the importance of external segregating factors. Clearly in 1961, segregation was considerably lower for the Irish and European Groups than for the two coloured minorities. By 1971, however, this position had been considerably modified. A decline in the numerical importance of the Irish and European born communities is reflected in a notable fall in the degree of residential segregation. What is more remarkable is the 13.3% decline in the segregation of the West Indian community at a time when the size of this group more than doubled. Although, recording a similar numerical increase, the Asian born community was characterised by an increase in segregation. Thus while Table 9.5 has indicated the reality of external factors in influencing ethnic residential patterns, it was clearly not a factor operating equally for each immigrant group. Nor is a black versus white immigrant group operational differential identifiable. Rather it is proposed that it is the duration of an immigrant group residence in a society which influences the extent to which external factors of segregation are operationalised. Husain (1975) has demonstrated that the Irish, European and West Indian communities are far more established, in historical terms, in Nottingham than are the Asian. Familiarity with a minority, by the host society, would appear to remove the intensity and structural forces in producing residential segregation.

TABLE 9.4
Nottingham C.B.: Index of Dissimilarity—1971

	Birth Place	1	2	3	4
1.	Ireland	-			
2.	Asia	40	-		
3.	West Indies	37	44	-	
4.	Europe	25	40	44	-

TABLE 9.5
Nottingham C.B.: Index of Segregation

	Birth Place	1961	1971
1.	Ireland	38.2	30.5
2.	Asia	61.5	64.2
3.	West Indies	69.7	56.4
4.	Europe	40.5	33.0

CONCLUSIONS

This paper has demonstrated that, in Nottingham C.B. considerable variation in residential similarity were identifiable between the major immigrant groups in both 1961 and 1971. These variations remained stable and thus suggest that internal cultural differences have strongly influenced the residential distributions of particular ethnic groups. However, the importance of external segregating influences was also demonstrated. However, it was suggested that in the increasing length of residence of the minority group, these factors decline in importance. Thus whilst it is possible to find evidence to support aspects of both in the internal and external theories of segregation, the precise balance between these two sets of factors obviously varies between ethnic groups and their length of residence in the host community.

REFERENCES

Cater, J. and Jones, T. Ethnic Residential Space: the case of Asians in Bradford. **Tijdschrift voor Econ. en Soc. Geografie,** 70, (2), (1979) 86-97.

Dalton, M. and Seaman, J.M. The distribution of New Commonwealth immigrants in the London Borough of Ealing **Transactions of the Institute of British Geographers,** 58, (1973) 21-69.

Duncan, O.D. and Duncan, B. **The Negro Population of Chicago.** University of Chicago Press (1957).

Herbert, D.T. and Smith, D. M. **Social Problems and the City.** Oxford University Press, 158-185.

Husain, M.S. Residential and intra-urban mobility patterns of Asian immigrants in Greater Nottingham. Unpublished Ph.D. thesis, University of Nottingham (1975)a.

Husain, M.S. The increase and distribution of New Commonwealth immigrants in Greater Nottingham. **East Midland Geographer,** 6 (3), (1975)b 105-129.

Jones, P.A. Some aspects of the changing distribution of coloured immigrants in Birmingham 1961-1966. **Transactions of the Institute of British Geographers,** 500, (1970) 199-209.

________Coloured minorities in Birmingham, England. **Annals of the Association of American Geographers,** 66, (1976) 89-103.

________Ethnic areas in British Cities, (1979).

Jones, T. and McEvoy, D. **Race and space in cloud cukoo land.** Area 10, (1978) 162-166.

Lee, T., **Immigrants in London. New Community,** 2, (1973) 145-159.

Peach, G.C.K. Urban social segregation, Longman (1975). Race **and Space,** Area 11, (1979) 82-84.

Peach, G.C.K. and Winchester, S.W.C. Birth Place, ethnicity and the enumeration of West Indians, Indians, Indians and Pakistanis in the censuses of 1961, 1966 and 1971. **New Community,** 3, (1974) 20-29.

Woods, R.I. Aspects of the scale problem in the calculation of segregation indices. **Tijdschrift voor en Soc. Geografie,** 67, (3), (1976) 169-174.

TEN

Elisabeth Lichtenberger

The Socio-Ecological Structure and Spatial Differentiation of Vienna in the 16th Century

THE SCIENTIFIC PROBLEM

In the English-speaking countries, factorial ecology and social area analysis more or less lay claim to the field of urban research, especially the spatial organization of urban society. The author's approach is a different one. In order to prevent misunderstandings it seems, therefore, necessary to define (1) the assumptions implicit in those research approaches; and (2) the author's own access to the problem.

ad (1): (i) Both factorial ecology and social area analysis form quite fascinating "mesalliances" between the biological determinism of association and succession of certain species within a specific territory and the economic location theory based upon the market mechanism of supply and demand and determined by land value.

(ii) Both are offsprings of North-American conditions, based upon politiical liberalism and private capitalism.

(iii) A historical dimension is lacking in the location theory. The study of historical phenomena is tied to the biologic concept of succession only.

(iv) There is no reflection on the aggregation of the data into spatial units. Fixed statistical units (census districts, wards, etc.) form the backbone of the analysis. With the diffusion of factorial ecology and social area analysis from North-America into foreign countries the problem of the statistical basis was even more aggravated. Moreover, the methods of data collection and processing applied differ widely. For this very reason the results cannot be compared.

ad (2): The author's approach differs basically both in theory and methods: (i) A cultural stage model is being used as a theoretical framework. Two ideal types of cities are defined,

the mediaeval burghers' town and
the baroque residence town.

They were the result of two political systems, the mediaeval feudal territorial state on the one hand and the absolutistic centralized state on the other hand, tied up with two different social systems.

Thorough research was centred on the problem in what way the mediaeval burghers' town was transformed into the baroque residence town.

(ii) A growth model formed the background against which political and social changes have to be seen. At the core of the study there is the analysis of the compulsory quartering of the court personnel into the burghers' houses, an outcome of the political subordination of the city council under the state bureaucracy. The result was a partial displacement of the population and the economic activities from the town into newly developed suburbs.

(iii) The methodology of the research project is based upon a classification system of house types, using the theoretical concept of M.Weber's ideal types (cf. below).

(iv) These house types constituted the level of aggregation of the following data: size of dwellings, number of storeys, house owners, number of stores, wine cellars, stables, occupation of both the house owner and his tenants.

THE SOURCES AVAILABLE

Important political changes sometimes provide interesting sources in connection with changing requirements. Accordingly the settlement of the Hapsburgs in Vienna in 1526 and the necessity to house the court personnel with the burghers of the town brought about the establishment of a court quartermaster's office ("Hofquartieramt") the main task of which consisted in creating a bureaucratic procedure controlling this quartering. For this purpose, court quartermaster's books ("Hofquartierbucher") were set up. The very first one dates back to 1563. It comprises a survey of the citizens' houses, their owners and the tenants. Though there were similar catalogues both in the Spanish and the Dutch bureaucracies before, this one still has

FIG. 10.1

DIE WIENER ALTSTADT
INNERSTÄDTISCHE PENDELWANDERUNG DER BEAMTEN 1563

KARTE 7
Entwurf: E. Lichtenberger
Zeichnung: H. Petritsch

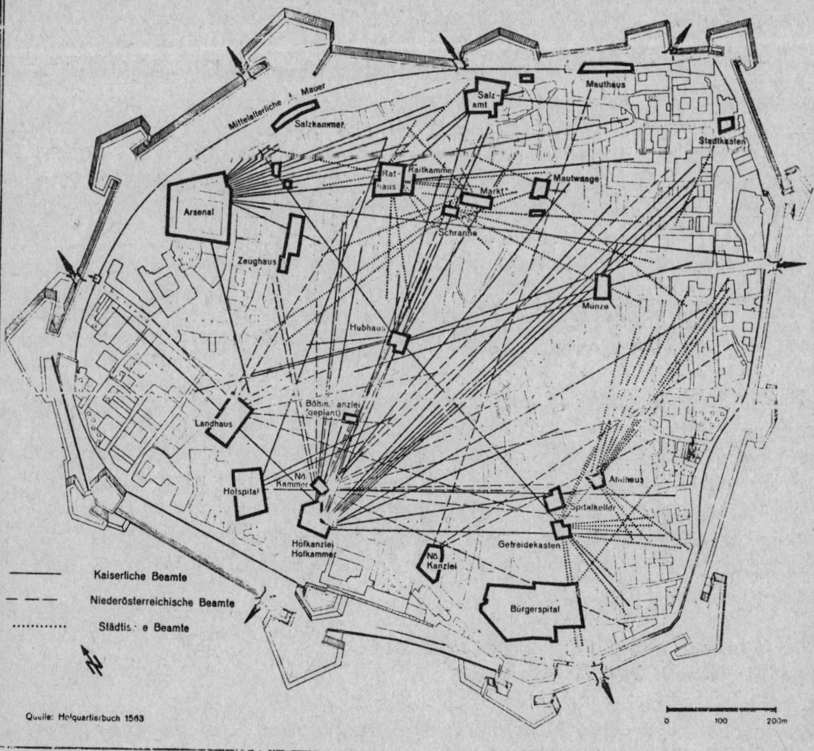

the directness and liveliness so very characteristic of the earliest products of bureaucratic institutions that are not as perfectly formalized yet as the later ones. Notwithstanding the fact that it has been known for some time, it has not attracted the attention of historians. Due to the fact that some houses are missing and proper names alternate with trades or professions it has the reputation of being inaccurate, but it is the only source available up to the early 19th century giving the occupations of the tenants and, therefore, rendering an almost complete social analysis of the "bourgeoisie" possible.

With the occupational structure the residential structure of each household was registered. The combination of these two sets of data appears the more important because of the fact that only the most recent official statistics (1971) allow for a similar one by means of computer facilities.

In connection with the second "Hofquartierbucher", dating back to 1566, it is possible to give detailed information on the building structure of the town (cf. above).

In analysing these data two methods were used:

(i) a statistical analysis; and
(ii) a cartographical representation.

In this context a difficult problem arose and had to be tackled, namely that of the localization of the data. The very first numbering of the houses recorded on a map dates back to 1779 only. Therefore, a rather complicated and time-consuming retrospective technique had to be applied to ensure a house by house allocation of the data of the mid-16th century. For the cartographic representation Daniel Suttinger's map of 1683 provided the basis.

The Spatial and Social Structure of the Mediaeval Burghers' Town

As early as in the 13th century Vienna was one of the large cities of the German-speaking countries. Even then it was characterized by a dichotomy of burghers and nobility.

The dukes of the Babenberger dynasty had laid the foundations for this by making Vienna the site of their castle in 1155/56. Since then the town had remained a residence, with intervals though. Therefore, Vienna can be considered the prototype for the development of European residence towns, which differ fundamentally both in their socio-economic structure and development from the "Freie Reichsstadte" so very important in the German-speaking countries which tended to form political and commercial leagues, later on, like the Hanse.

Similar to other towns situated on the former Rhine and Danube-lines - the boundary of the Roman Empire - mediaeval Vienna came into existence within the walls of the Roman castrum by the settling of merchants engaged in long-distance trade (Wik) close to a castle towards the end of the 10th century (Fig.10.2).

As early as in the 11th century a planned suburb of foreign merchants was formed at the long-distance road to the east. This determined the location of wholesale for a number of centuries.

Next to the palace of the Babenbergers the ghetto was erected on the ducal estate soon after 1200 and remained there until 1421.

The large extension of the town early in the 13th century enclosed the merchants' suburb mentioned above as well as the ecclesiastical quarter around St. Stephen's cathedral that had formed about 1137 and provided room for new social groups and institutions, such as large monasteries and hospitals on the meadows at the periphery of the older centre of the town.

In connection with the ducal palace being removed from the city core to the periphery a quarter of the nobility developed during the 14th century. Attempts at giving it a special legal status as a "town of the nobility" failed, when Rudolf IV declared the land taxes dischargeable and granted permission to acquire urban lots to all burghers, even the artisans, as from the mid-14th century. At the same time the university was founded (1365) — now the oldest one in the German-speaking countries — giving rise to a college quarter (Fig. 10.3).

Presumably under the influence of the residence function, at an early date a rent-capitalistic spirit developed in the patriciate, whose leading members attained the status of noblemen and invaded the sphere of the lords of the manors. They did participate in long-distance trade, but normally left the risk to foreign merchants (mainly from the areas along the Danube and the Rhine) and concentrated on the attractive branch of commission business privileged by the sovereigns. Similarly in request there was the tenure of public offices which definitely was of a speculative character and invested the creditors of the sovereigns with considerable political power.

1G. 10.2 The development of Vienna in the Middle Ages. Interior oldtown.

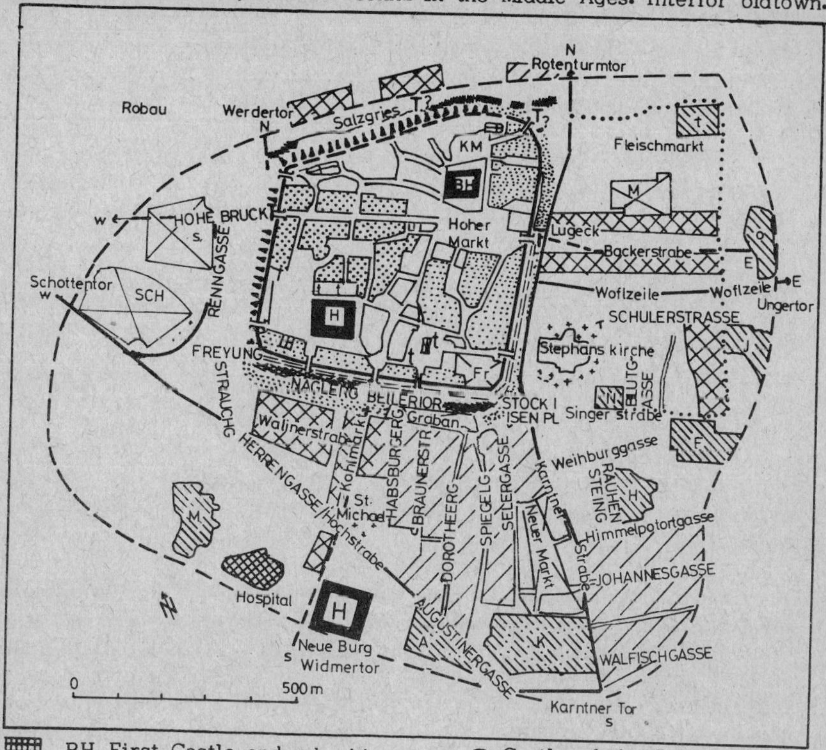

BH First Castle and suburbium B Castle of the Babenberger

Wik

Ghetto

former Intervallum of the Roman castrum

Fortification

=▼ Ditch resp. slope

T Old gates

Mediaeval extension of the town about 1200

former suburbs H Castle of the hopsburgs

Earth ditch (?) at the middle of the 12th century

gate streets

Planned layout

squares

street blocks

Peripheral real estate

ecclesiastical monor-house

monasteries

hospitals

cemeteries

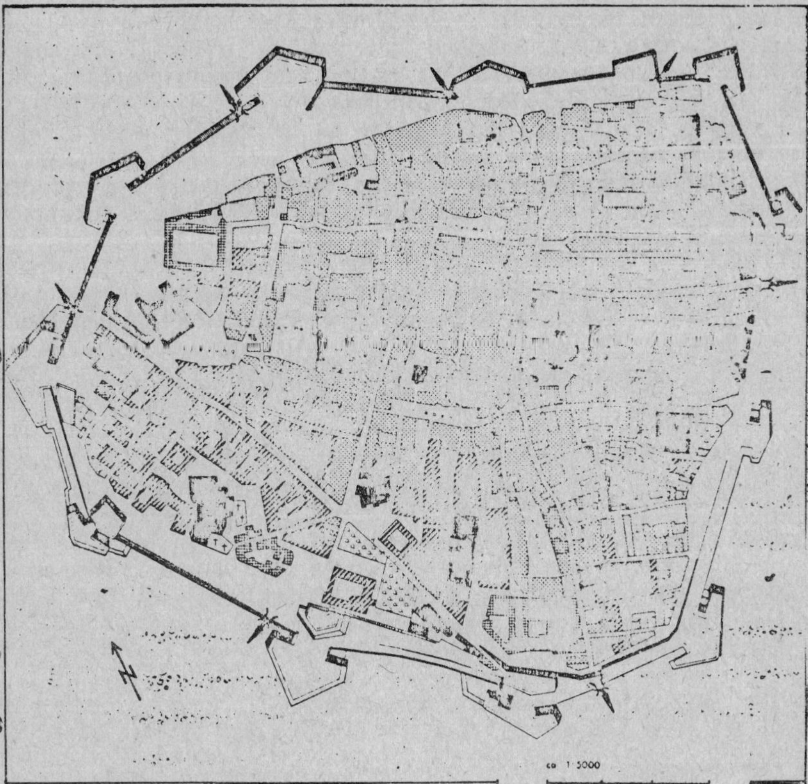

FIG. 10.3 The building structure 1566

- Public buildings
- Churches and monasteries
- Ecclesiastical manor houses
- Houses of the nobility
- Burgher's houses - three stories and more
- one and two stories
- Barn
- Garden

The possession of houses and vineyards in the surroundings of Vienna, the leasing of shops, store-rooms and dwellings obviously was considered more important than the development of a prosperous export trade.

Altogether, therefore, export trade remained comparatively insignificant during all of the Middle Ages compared with the diversificiation and abundance of consumer trades. Accordingly the putting-out system did not attain great importance. Thus the town was spared the social conflicts which arose in the export towns of Continental Europe late in the Middle Ages. There were no masters exploited by entrepreneurs, and no large body of journeymen gave rise to riots as the average number of journeymen of the individual masters in Vienna was surprisingly low and exceeded two in rare cases only.

In general class boundaries as between patricians and artisans were permeable. Especially the members of the highly esteemed luxury goods and food trades, like butchers and furriers, were connected with the patricians due to

(1) their considerable share in real estate, and
(2) participation in the long-distance trade by means of which they were enabled to rise to the leading group.

This trend to acquire real estate turned out to be a severe handicap in the residence town of the 16th century, when new parties became interested in landed property. It was too late, then, for the burghers' upper classes to make export activities the basis of their economic support.

The Political and Economic Framework of the 16th Century

Vienna entered the 16th century as a big but impoverished town. The "Central German Road" ("Mitteldeutsche StraBe") had outdone the Danube line, and the Osman invasion of Southeastern Europe had robbed the town of most of its outlet for export goods. Burghers, upper and middle classes had become poor. The guilds had been closed, i.e., the numbers of masters and journeymen were limited. Only viticulture seems to have been able to hold out in this general decline. Therefore, one could speak of "re-agrarization" of the burghers early in the 16th century.

Being implicated into the struggle for power of the nobility with the sovereigns, the town had to atone for it in the Wiener Neustadt Trial ("Wr. Neustadter Blutgericht") of 1526. The members of the patriciate lost both their property and their lives. The new charter of the town sealed the loss of the former burghers' autonomy and brought about a complete subordination of the town authorities under state administration.

After the first Osman siege of 1529 Vienna became a border-town and fortress and remained in this fairly precarious situation for one and a half centuries. In this way the development of suburbs was prevented, involving a competition for the scarce urban space after the Hapsburgs — during the reign of Ferdinand I — had decided to make Vienna their residence town again after a one hundred years' interval. Having attained great importance in the European political scene meanwhile, they started to develop an empire in Europe's southeast from here.

Architectural Changes in the Town and the Effects of the Compulsory Quartering of Court Personnel

On looking back it seems to be of extraordinary significance that this remodelling of a mediaeval burghers' town into a baroque residence took place without any more substantial architectural interference on the part of the Hapsburgs. A number of reasons may be given for this:

(i) Because of the confrontation of the Austrian monarchy with the Osman Empire urban design and building activities concentrated on the fortifications of the border towns, especially of Vienna. After the second siege of Vienna in 1683 and the victorious war against the Osman Empire it was the resettlement of the devastated Great Hungarian Plain that not only attracted immigration but also depleted the public treasury.

(ii) The ownership of real estate undoubtedly played an important role as the Hapsburgs had sold their urban property early in the 14th century but for a few lots, whereas, e.g., in Berlin the Hohenzollern had a quite different policy. There in the late 18th century Frederic the Great owned forty per cent of the urban area, which offered the basis for a remodelling of the town.

(iii) The architectural interests of the individual member of the Hapsburg dynasty—which, as such, was a highly artistic and gifted line of rulers—were limited to the erection of a few buildings. They had no intention to remodel the capital of their empire in the same way as the French kings did in Paris and, later on, the Hohenzollern in Berlin.

(iv) The housing of the numerous court employees was managed by means of the traditional instrument of compulsory quartering. This system of unwished-for tenants in the burghers' houses was abolished as late as in the 18th century.

The effects of this compulsory quartering were multifarious: It is obvious that the houseowners burdened with it offered resistance and were determined to evade it. Thus we read in the court quartermaster's books of locked and walled up rooms, devastated ones, where window-frames and casements had been found lying on the floor.

But there were positive effects, too. Because the court quartering office offered "tax free years" in case of renovation or rebuilding, this was a considerable incentive for extra building activities and one of the main reasons for the burghers' participation in the baroque remodelling of the town.

In spite of all the eloquent complaints of the burghers of considerable damage caused by the court personnel, there definitely was an undeniable economic improvement due to the increasing immigration of well-to-do consumers. The decisive socio-psychological factor consisted, however, in the confrontation of the social strata of the burghers' elite with the nobility, the artisans with the court personnel. For good reasons it can be said of the baroque building activities that the burghers built in the manner of the nobility, just as they tried to imitate its way of life.

Due to the growing demands on housing space of the highly differentiated court, of an increasingly complex civil service and the nobility the most impotent group in the social structure, namely the artisans' class, had to give way. This situation is mirrored in the marginal notes on the handicraftsmen in the court quartermaster's book which read: "May be given notice", "May reside at a greater distance from the court", "May live outside the town" or even "May be expelled". Some of the artisans made an effort to stem these tides by pointing to the fact that they had been tenants in these houses for twenty or more years.

The first court quartermaster's book (1563) shows the start of the invasion of the new court-oriented classes, and thus of a process extending over more than 200 years until the mid - 18th century when the middle classes had lost most of their poossessions in real estate within the town to the highly differentiated nobility and the clergy.

House Types and Social Differentiation in the Mid-16th Century

Table 10.1 summarizes the functional differentiation of the houses in 1566.

TABLE 10.1
The functional differentiation of the houses in Vienna in 1566

house types	number of houses	per cent
burghers' house	925	81,9
noblemen's houses	59	5,2
houses of the clergy* and church	87	7.7
public buildings		
of the town	28	2,5
of the country	16	1,4
of the court	2	0,2
university dormitories	12	1,1
	1129	100,0

* monasteries, convents, churches, buildings owned by monasteries domiciled outside Vienna, beneficiaries' houses.

The comparatively small number of noblemen's houses veils the fact that, because of the larger size of their lots, about one sixth of the town's area was owned by the nobility. With respect to the property of the clergy it is the houses owned by religious bodies domiciled outside the town that are of greatest interest. Partly they were possessed by potent bishoprics, such as Salzburg, Passau, Trient, partly by large monasteries owning land in the surroundings of Vienna. They served their purpose as lodging houses and for celebrations. Furthermore some of them constituted a sort of dairy-farm in town. Even at an early date tenants were accepted into these houses of the clergy.

With respect to the types of burghers' houses the views presented in the literature so far were inadequate. Many authors arrived at false conclusions in paralleling each of the gable-houses of old panoramas with one lot and thus with one owner. As the result of a reconstruction of the network of plots in Vienna in the 16th century and an analysis of old house plans the following types of houses could be distinguished:

(i) The gable houses mentioned above with a frontage of

two or three windows towards the street occupied deep and narrow lots and had originally been owned by artisans. Later on they were turned into tenement structures by adding storeys and combining two (or more) of them (Fig.10.4).

(ii) On the other hand there were spacious houses arranged around large courtyards ("Wohnhof") on lots of 10 or 20 times the size of those mentioned above. There is proof of many fortress-type towers—corresponding to the dwelling towers in Italian towns—as well as of chapels. Dwellings in these large houses—owned by the Vienna patricians—had been leased since the later part of the Middle Ages (Fig.10.5).

(iii) According to the importance of viticulture for the Vienna economy a specific urbanized farm-type house existed, situated on comparatively wide lots, with doorways for the passage of vehicles and long sidewings, in which there was room for stables, too, as opposed to the situation in the tradesmen's houses (Fig.10.6).

(iv) In areas near the edge of the town there were small houses on tiny lots.

Last but not least it should be mentioned that the outline of blocks as well as structural elements of some of the former craftsmen's houses still exist today, whereas, according to the basic principles of urban development, the objects on the large lots have undergone reconstruction, occasionally several times, since the 16th century.

The Extent and the Owbers of Tenement Structures in the 16th Century

The notion of the burgher as a house-owner is among the cliches associated with preindustrial European urbanism. As can be gathered from certain facts mentioned, it was no longer true in the Vienna of the late Middle-Ages. As far as the period analysed in this paper is concerned it must be completely revised.

When adding up the individual items on owners and tenants in the court quartermaster's book, it becomes obvious that in 1563 already 70 per cent of the dwellings in the burghers' houses were let.

From table 10.2 it can be inferred that almost two thirds of the tenants lived in houses with 3 or more tenants and one third in houses with more than 5 dwellings. Only one fith of the

FIG. 10.4

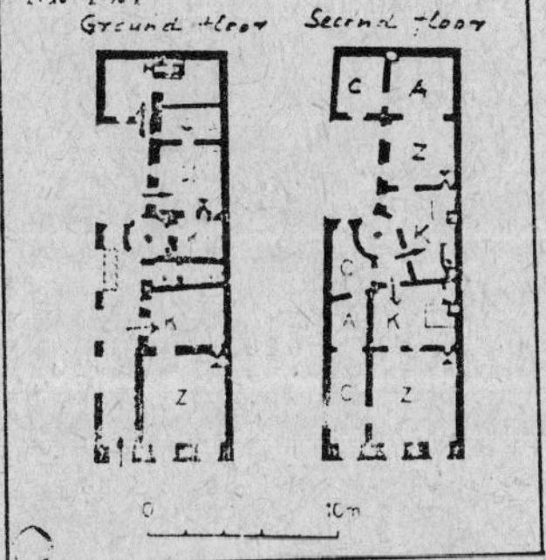

FIG. 10.5

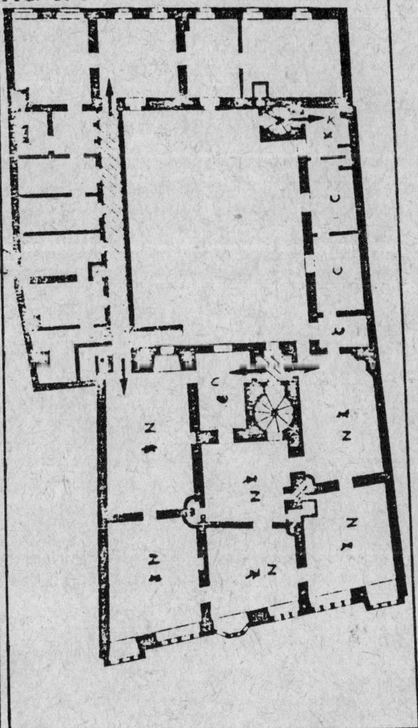

FIG. 10.6

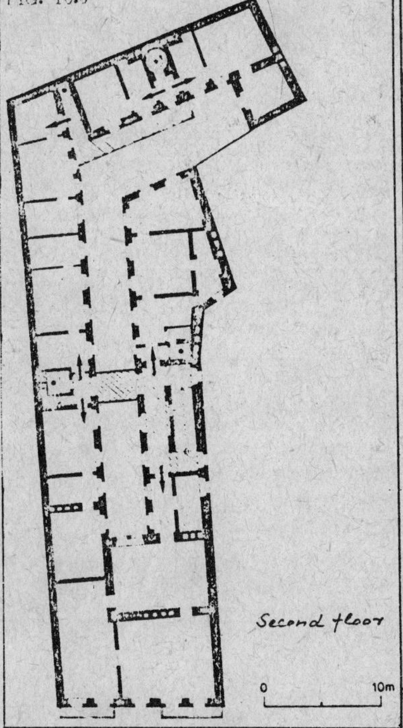

Mediaeval gable house of tradesmen

→	Dwelling entrance		Chimney
	Corridor, stairhouse	A	dark room
V	entrance room		Open corridor
Z	room (two windows)	K	Kitchen
C	Closet (one window)	⊡	W.C.

Courtyard-tenement house of patricians

Second floor

Baroque remodelling of an urbanized farm-type house

burghers' houses had no tenants. Another interesting item presented in this table is the fact that only 817 houses were lived in by their owners as well, whereas 148 can be classified as "second houses". When comparing the latter with the houses lived in by their owners, it can be deduced, that most of them were considered as an investment of funds, as the number of tenants was considerably larger.

TABLE 10.2

The number of tenants in Vienna's burgher houses in 1563

Number of tenants per house	Main and second houses		House owners		Tenants		Inhabited dwellings	
	number	%	number	%	number	%	number	%
0	197	20.4	185	22.6	-	-	185	6.8
1	292	30.3	249	30.5	292	15.3	541	19.9
	210	21.8	175	21.4	420	22.0	595	21.9
	110	11.4	99	12.1	330	17.3	429	15.8
	63	6.5	43	5.3	252	13.2	295	10.8
	43	4.5	32	3.9	215	11.3	24	9.1
	13	1.3	7	0.9	78	4.1	85	3.1
/	15	1.6	13	1.6	105	5.5	118	4.3
8	9	0.9	7	0.9	72	3.8	79	2.9
9	5	0.5	3	0.4	45	2.4	48	1.8
10	3	0.3	1	0.1	30	1.6	48	1.8
11	3	0.3	1	0.1	33	1.8	34	1.2
14	1	0.1	1	0.1	14	0.7	15	0.5
20	1	0.1	1	0.1	20	1.0	21	0.8
	965	100.0	817	100.0	1906	100.0	2723	100.0

An analysis of the houses that were lived in, apart from their owners, by quartered court personnel only provides an insight into the older structure of the tenant system in Vienna as it had existed before the revival of its residence function. It becomes obvious that it effected mainly houses formerly without any tenants owned by artisans, into which one or two court employees were quartered compulsorily.

In the literature frequently the opinion is expressed that the taking in of tenants started by letting whole storeys to one tenant. This holds true for Vienna in rare cases only.

The compulsory quartering undoubtedly brought about a partitioning-off of small flats, and, moreover, an increase of the housing density. The generally extremely modest housing conditions in the burgher houses can be gathered from table 10.5:

TABLE 10.3
Differentiation of the burghers' main and second houses according to the number of tenants in 1563

Number of tenants por houco	Main houses number	%	Second houses number	%
0	185	22.6	12	8.1
1	249	30.5	43	29.1
2	175	21.4	35	23.7
3	99	12.1	11	7.4
4	43	5.3	20	13.5
5	32	3.9	11	7.4
6 and more	34	4.2	16	10.8
	817	100.0	148	100.0

TABLE 10.4
The tenants' structure of the burgher houses before the compulsory quartering of 1563

Number of tenants per per house	Main houses number	%
0	299	36.6
1	153	18.7
2	161	19.7
3	98	12.0
4	41	5.0
5	31	3.8
6 and more	34	4.2
	817	100.0

This table shows that

i the lodging-conditions of the house-owners were, on an average, better than those of the tenants and
ii more than 70 per cent of the tenants had to put up with very small lodgings with one or two rooms only.

TABLE 10.5

The size of the dwellings of house-owners and tenants in Vienna's burgher houses in 1563

Size of dwellings	House owner		Tenant		Total	
	number	%	number	%	number	%
Kitchen + 1 room (closet)	1	0,1	314	16,5	315	11,6
Kitchen + 1 room + 1 closet	201	24,6	1025	53,8	1226	45,0
Kitchen + 2 rooms (or 1 room + 2 cl.)	215	26,3	343	18,0	558	20,5
Kitchen + 2 rooms + 1 cl. (or larger)	376	46,0	181	9,5	557	20,5
No information	24	3,0	43	2,2	67	2,4
	817	100,0	1906	100,0	2723	100,0

This problem of a housing shortage combined with the phenomenon of small flats can be observed in Vienna since then through the centuries to the present, changing, however, due to the growth of the town, from a problem of the core area into one of the respective newly added quarters. Only the present succeeded in effecting a decisive improvement.

Among the owners of the tenement-houses the patricians occupied first rank, housing about one half of the tenants, with a tenants-owner ratio of 4.1 (1.6 with the artisans). As to the buildings the spacious units arranged around courtyards mentioned above were predominant.

It should be stressed that the intellectual elite, appearing as "Doctoren" and "Magister" in the court quartermaster's book, though small in number only, had the highest ratio of tenants (5,8).

Whereas the statistics of the house-types still reflect the proportions of the mediaeval burghers' town, the analysis of the social statistics (cf. Table 10.6) already mirrors the beginning of the residence period.

The table shows that only about 60 per cent of the dwellings were lived in by burghers (in sensu stricts) whereas as many as 40 per cent were occupied by the nobility, court personnel and officials.

No statements can be made about the total population, as no information is available on the numbers of the clergy, the hands living with the artisans families, the servants as well as on vagrants, street-players, vagabonds, beggars and prostitutes.

The house owners' ratio gives the percentage of house owners of the total number of owners and tenants. It shows that both the patricians as well as the intermediate group of the wine-burghers normally possessed houses of their own. There is a gap between them and all the other social groups. In this context it should be noted that the houseowners' ratio of the artisans (33,2) was almost identical with that of the nobility, of which a comparatively high percentage lived in rented lodgings due to the considerable amount of immigration. A complex group was formed by the developing bureaucracy. The town officials were closely connected with the domiciled burgher elite and, therefore, participated in the house-ownership, whereas the officials of both the court and the state were among the large number of tenants just as the partly foreign court-society, whose leading members, however, succeeded

TABLE 10.6
The social structure of house-owners and tenants 1563

	Social Structure	House owner number	House owner %	Tenant number	Tenant %	Total number	Total %	Quota of house-owners
(1)	Agriculture	7	0,8	25	1,3	32	1,1	21,8
(2)	Burghers							
(a)	Propertied burghers	181	20,7	35	1,8	216	7,5	83,8
(b)	Vine burghers	71	8,1	-	-	71	2,5	100,0
	Patricians	252	28,8	35	1,8	287	10,0	87,8
(c)	Burghers engaged in handicraft trades etc.							
	Artisans	320	36,5	645	32,2	965	33,6	33,2
	Services	21	2,4	56	2,8	77	2,7	27,3
	Trade	62	7,1	172	8,6	234	8,1	26,5
	Transport	7	0,8	26	1,3	33	1,1	21,2
	Catering	37	4,2	68	3,4	105	3,6	35,2
	Sum 2 (c)	448	51,0	967	48,3	1415	49,2	31,7
(d)	Widows etc.	1	1,1	82	4,1	83	2,9	1,2
	Sum 2	700	79,9	1084	54,2	1784	62,1	39,2
(3)	Employees at the Court							
	Court Servants	17	1,9	247	12,3	264	9,2	6,4
	Guards	6	0,7	81	4,0	87	3,0	6,9
	Sentinels	1	0,1	95	4,8	96	3,3	1,0
	Singers and musicians	4	0,5	59	3,0	63	2,2	6,3
	Sum 3	28	3,2	482	24,1	510	17,7	5,5
(4)	Administration and University							
	Officials	65	7,5	263	13,2	328	11,4	19,8
	"Doctors" and "masters"	17	1,9	26	1,3	43	1,5	39,5
	Sum 4	82	9,4	289	14,5	371	12,9	22,1
(5)	Nobility	59	6,7	119	5,9	178	6,2	33,1
	Sum 3 - 5	169	19,3	890	44,5	1059	36,8	16,0
	Total	876	100,0	1999	100,0	2875	100,0	30,5

in acquiring houses soon.

THE INTERMINGLING OF THE SOCIAL GROUPS

The question is to be posed to what extent an integration of the social groups was brought about by the tenement house system and the compulsory quartering. Who was accepted as a tenant by the patricians, by the officials, by the court servants and by the artisans? A matrix of the data offers the following results:

(i) The burghers engaged in trades and commerce (artisans, members of the services, trades, traffic and catering) constituted about one half of all the tenants and were distributed almost equally over all groups of house-owning burghers. It should be mentioned in this context that up to the late 19th century this social stratum always had a very low segregation index when analysed as whole. Independent of this general trend there were, however, very marked associations of specific branches.

(ii) Only the group of the widows was, due to their, on an average, very meagre incomes, overrepresented in the artisans' houses. Exactly the opposite was true of the small number of patricians among the tenants. They normally found lodgings within their own group or in the larger houses of the members of traffic and catering trades or of town officials.

(iii) The trends towards segregation differed widely with the new group of tenants constituted by the court servants, the state officials and the nobility. The segregation was practically insignificant with the civil servants, who were, however, a highly differentiated group. There was a certain affinity towards house—owning officials. Besides renting lodgings in burghers' houses, civil servants were provided with housing in public buildings as well as in houses owned by monasteries and similar clerical institutions.

(iv) The distribution of the court servants depended on the quartering policy, which tended to take rather drastic measures towards the artisans who were no longer represented in the town council as from 1526 (cf. above). That's why they had to house a by far larger proportion of the court servants than the other groups of the burghers. Patricians, officials and court attendants had a better chance to protect themselves against the quartering of these—very often not well-off—court servants.

(v) The big and representative rooms of the spacious tenement houses of the burghers' elite—that were often described as "halls" in the court quartermaster's books—were the favourite lodgings of the nobility that "flooded" Vienna in increasing numbers. No wonder that, accelerated by the strong mobillity of the ownership of houses, the nobility became the owners of the large patricians' houses within a few decades. An important feature in connection with the quartering of the nobility was the possibility of accommodating horses. On account of the interior structure of house-types it can be stated that, in principle, the existence of stables was limited to the houses of those burghers, who could afford horses ("die Rosse vermogen"), i.e., the large tenement-units around courtyards and the urbanized farm-type houses.

Though it is impossible to differentiate between the keeping of horses because of economic necessity or social status by means of the data given in the court quartermaster's book, an estimate made on the basis of the stables registered in 1563 amounts to a capacity of about 4000 horses that could be accommodated in the stables of the houses of both the burghers and the clergy. There is no information available on the nobility's horses.

Altogether it can be said that the result of the quartering system was a marked intermingling of social groups, if the individual house is taken as the object of study, An especially close contact is to be observed between

(i) the nobility and the burghers' elite, and
(ii) tho artioano and court ocrvanto.

In this way, often members of different social status lived under the same roof.

THE BAROQUE REMODELLING OF THE TOWN AND THE SEGREGATION OF THE SOCIAL GROUPS

Only in the course of the baroque remodelling of the town during the 17th and early 18th centuries a more pronounced segregation of the social groups took place, i.e., the baroque building activities brought about a new identity of building type, ownership and social structure.

Some statements may serve as an illustration:

(i) The mediaeval artisans' houses offered mostn resistence against remodelling tendencies - one or two storeys were added or two or several of them put together.

TABLE 10.7
The intermingling of the social groups in 1563

House owner	Tenants in burgher houses																	
	Patricians		Artisans		Court personnel		Officials		"Doctoren"		Widows		Noblemen		Others		Total	
	number	%	number	%	number	%	number	%	number	%	number	%	number	%	number	%	number	%
Patricians	16	47,2	354	38,5	153	33,7	81	35,4	7	41,2	27	35,1	66	62,[illegible]	32	45,7	736	38,6
Vine-burghers	1	2,9	61	6,6	33	7,3	25	10,9	-	-	6	7,8	4	3,8	3	4,3	133	7,0
Artisans, services and agriculture	4	11,8	247	26,9	173	38,0	62	27,1	3	17,6	26	33,7	11	10,4	14	20,0	540	28,4
Trade	3	8,8	65	7,1	34	7,5	13	5,7	-	-	1	1,3	7	6,6	3	4,3	126	6,6
Traffic and Catering	5	14,7	48	5,2	18	4,0	15	6,5	-	-	3	3,9	3	2,8	4	5,7	96	5,0
Court personnel	1	2,9	25	2,7	8	1,8	4	1,7	-	-	2	2,6	3	2,8	2	2,9	45	2,4
Officials	3	8,8	64	7,0	21	4,6	21	9,2	3	17,6	5	6,5	10	9,4	5	7,1	132	6,9
"Doctoren"	1	2,9	55	6,0	14	3,1	8	3,5	4	23,6	7	9,1	2	1,9	7	10,0	98	5,1
	34	100,0	919	100,0	454	100,0	229	100,0	17	100,0	77	100,0	106	100,0	70	100,0	1906	100,0

Tradesmen, officials of low rank and court servants constituted the tenants.

(ii) The agrarian burghers' houses had to give way to monasteries, palaces and tenement structures (Fig.10.5)

(iii) The courtyard-tenement houses of the patricians were replaced by

- palaces of the nobility either for the use of the individual noble families or for being let to foreign merchants, officials of high rank and other members of the ennobled burghers' class,

- large baroque tenement houses with a marked mingling of social groups.

(iv) The clerical manor houses became huge tenement structures, especially for officials of high rank and the members of the intellectual elite of the town.

(v) The tiny houses on the bastions lodged the members of the town guard and other low class people.

In addition to the described social segregation according to house types baroque remodelling brought about a considerable segregation within the individual houses. There was a social differentiation due to the fact that accommodation in the front building and on the first floor was rated higher than lodgings in the backyards and on the floor higher up. As a general principle the front building was considered more exclusive, and, therefore, always contained larger apartments. It also had, however, a vertical differentiation as to the size of apartments and social structure: Normally the largest apartment—that of the house-owner—was situated on the first floor, and not only the height of the storeys but the size of the apartments and, accordingly the social status of the inhabitants decreased upwards.

The "class-tied" demographic structure

While demographic and social segregation became more or less independent phenomena in the modern city of the western industrial society the situation was entirely different in preindustrial towns on the European continent, where there existed a "class-tied" demographic behaviour and, therefore, a specific demographic structure of the households. This is being discussed below on the basis of the conditions in Vienna in the 16th century (which remained more or less the same up to the middle of the 19th century!).

It was a common feature of the households of both the nobility and the burghers that they contained a large proportion of non-family-members and of, by definition, "non-resident" population as well, though there were great differences as to the composition and size dependent on status and property.

(i) A complex structure and very large households were characteristic of the nobility after it had become resident and had been provided with its own palaces. Usually a three-generation-unit of an extended family including several married couples and some more distantly or closely related unmarried persons was combined with a body of servants, whose number amounted to a multiple of that of the family members, and whose composition resembled that at the imperial court, on a smaller scale though. A few higher-ranking servants could even be married. In the palaces of the high nobility 100 and more persons used to live in such a large household during the baroque period. With regard to the number of children per family the nobility surpassed all the other classes.

(ii) In general the members of the guilds were married, a small percentage widowed. The protective measures for safeguarding the existence of the individual crafts were still in operation. The closure of the guilds made it a rule, that a master's widow had to remarry after a year of mourning. In this way the journey-men's marriage into a business was an important social institution. Due to the large differences in the ages of the partners there often was a small number of children only or even none at all.

(iii) The officials, at least those of the state and the court, soon differed from the artisans in their generative behaviour. For them, the existence of a housewife was not absolute must, whereas with the tradesmen their wives were indispensible for a number of functions, such as the catering for the apprentices and journeymen, help with serving the customers etc.

Moreover the officials received regular (and even then only meagre) salaries as from a fairly late date only, namely since the 18th century. In general a larger proportion of them remained single and in the case of marriage the number of children normally was smaller than with the artisans and the nobility. The officials were first in starting birth-control in the baroque period (!).

Principles of ethnic segregation did not gain any importance with the nobility. Language and style of living were dictated by court-etiquette. Only the group of the foreign merchants remained separate during the centuries discussed.

The separation of housing and working: the rise of the intra-urban commuting system

The separation of housing and working is generally considered a feature of the industrial period only. In Vienna it proved, however, an important urban phenomenon of the 16th century already and was clearly recognizable with three classes:

(i) In the nodes of intra-city traffic shops were packed closely together in shopping streets, as we learn from the court quarter-master's book, which gives, for a considerable number of houses, a number of four and more shopes in 1566. We know that the letting of shops was quite common in Vienna since the 14th century. Unfortunately the information given by the court quartermaster's book is insufficient for a precise separation between shops, workshops and dwellings. Therefore only an estimate can be made, based upon the surplus of shops over the apartments in a house, that at least 20 per cent of the tradesmen had already effected a separation of housing and working.

(ii) As early as in the Middle Ages a rather diversified administrative organization had developed corresponding to the tasks of town-government. Soon the organizations of the Lower Austrian government were added. As from 1526 the development of state government started under the rule of Ferdinand I. It is beyond the scope of this paper to discuss the organization of these authorities.

With regard to the ownerships there was a large range of houses and sizes of the apartments in all of these three groups according to the professional status which differed widely between the numerous copyists or clerks on the one hand and the holders of rather lucrative positions on the other hand.

Of the town-officials 40 per cent were house-owners. All the high ranking positions of the town administration were held by house-owning burghers of the patriciate.

Whenever civil servants of the provincial or the state government had houses of their own, and even when those were fairly large, they normally succeeded in evading quartering.

With the tenants, the bureaucratic hierarchy is clearly mirrored in the sizes of the dwellings, as very soon the supply was standardized according to different ranks.

The proportion of house-owners was by far smallest with the court officials. Even the top ranks, such as the head quartermaster himself (!), still had to live in rented dwellings in 1563. By means of an—in many cases fairly difficult—allocation of the individual officials to their places of work (which was possible in about one half of the cases) the intra-urban commuting system of the officials could be outlined for the mid-16th century (Fig. 10.7).

As to the spatial pattern within the town the officials held an intermediate position between the merchants and court-servants regarding traffic accessibility.

With respect to the distance of commuting moreover differences could be found between the town officials and those of the province and the state. Around the buildings of town authorities the respective officials lived in an average distance of 250 metres and none of them had to walk more than 500 metres. On the other hand, the officials of the province and the state had to cover about twice that distance, in a few cases even up to 1.5 kilometres.

(iii) The commuting of the court servants was even more important than that of the officials. As early as in the middle of the 16th century they constituted the second largest social group in town, outnumbered by the artisans only. There is no room here to discuss the structure of the court personnel in detail. It should be mentioned, however, that as early as in the middle of the 16th century there were already 91 different types of occupations.

As to the spatial distribution of the court servants a correlation between distance and rank in the bureaucratic hierarchy can be observed, subject to a central-peripheral principle of order varied by the influence of house types. The higher-ranking employees were housed in the centre, in the courtyard-houses of the patricians, the lower ranks at the periphery near the fortifications and in the artisans houses. A comparatively small number of court employees was to be found in the streets housing the town officials and of foreign merchants.

The "Hofburg" (imperial palace) was not only for too small for lodging the servants, it also did not have enough room for stabling the horses, therefore these had to be kept in the houses of the patricians and in the mansion-type houses of the

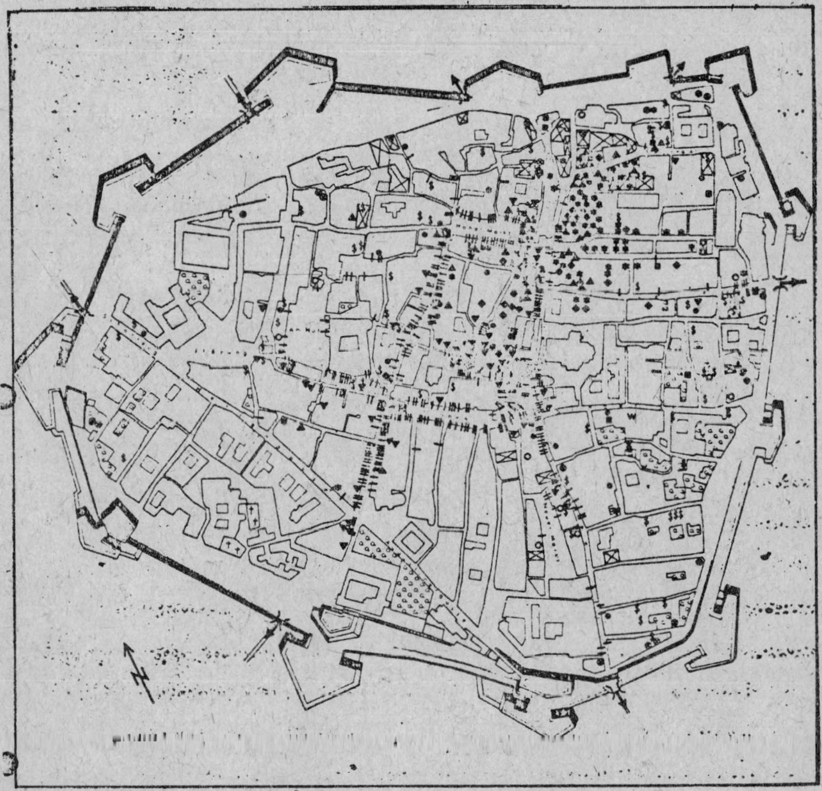

FIG. 10.7 Wholesale and retailing 1566

- Foreign merchants
- Viennese merchants
- Wholesale-dealer
- Grocer
- Other retailers
- Ironmonger
- W Money-changer
- $ Tavern-keeper
- Carriers
- fodder-dealer
- Shops
- Inns

clergy (cf. above). A synopsis of these three groups of commuters shows that 36 per cent of the Vienna heads of households had separated housing and working in the middle of the 16th century already.

With respect to the location of the places of work in the 16th century there was a marked tripartite structure, namely

(1) the Hapsburg residence
(2) the retail-city—both of those were centres of the intra-urban commuting at the same time—, and
(3) the quarter of foreign merchants and innkeepers along the Danube as the centre of export economy.

Sources:

1. Hofkammerarchiv Wien. Hofquartierbuch 1563, 1565, 1644, 1664.
2. Plan- und Schriftenkammer des Wiener Magistrats. Bauplane von Wohnhausern.

Maps:

1. Camesina A.v., 1880. Plan der Befestigungen und der Hohenverhaltnisse der Hauser der Stadt Wien im Jahre 1566. Wien.
2. Eisler M. (Hrsg.) 1919. Historischer Atlas des Wiener Stadtbildes. Wien.

REFERENCE

Lichtenberger Elisabeth, Die Wiener Altstadt. Von der mitterlalterlichen Burgerstadt zur City. Deuticke Wien, 410 p., Atlas (21 maps), Bibliography (1977).

Illustrations

(1) The development of Vienna during the Middle Ages
(2) The building structure 1566
(3) Mediaeval gable-house of tradesmen
(4) Courtyard-tenement house of patricians
(5) Baroque remodelling of an urbanized farm-type house
(6) Wholesale and retailing 1566
(7) Intra-urban commuting system of officials 1563
State officials
Officials of the provincial government
Town officials

ELEVEN

Gill C. Lim

Residential Segregation
An Empirical Analysis of Intercity Variation

INTRODUCTION

Despite numerous studies on racial residential segregation, the relative importance of various factors determining variations in the level of segregation among cities has not been fully analysed. Relatively more research effort has been directed to explaining intra-urban patterns of segregation, and most of the existing studies rely heavily on 1960 and pre-1960 data. The persistent trends of residential segregation in American cities over the 1960-1970 period points to the need for an analysis of causes of segregation using more recent data and a comparison with the previous results.

This paper examines the determinants of residential segregation by race using cross-sectional data for 43 cities in 1970. It differs from previous studies of segregation on two major respects. First, the paper is concerned with an analysis of intercity variation in residential segregation. A series of empirical models are developed and estimated to assess the relative importance of economic and attitudinal factors affecting variations in the level of segregation among metropolitan areas. Second, it provides analyses of both black-white and nonwhite-white segregation. Two separate sets of models are evaluated to observe major similarities and differences between black-white and nonwhite-white segregation.

In the following section the trends in residential segregation between 1960 and 1970 are briefly reviewed, and relevant theoretical issues are discussed. A major effort in this section is given to a discussion of various hypotheses regarding racial segregation. The third section reports on the method and results of empirical analyses. The empirical analyses are conducted by estimating a series of regression equations. The last section summarizes the results of the analysis and discusses policy implications.

Trends and Theoretical Issues

Considerable social and institutional efforts have been made over the last two decades to reduce the residential segregation by race in American metropolitan areas. In particular, the 1960's was marked by an impressive record of public actions aimed at the reduction in levels of segregation. The non-discrimination principle in federally assisted programs was enacted as Title VI of the Civil Right Act of 1964, and the fair housing concept was instituted as Title VIII of the Civil Right Act of 1968. In the ensuing years fair housing laws were also established at lower levels of government.

Nevertheless, available evidence suggests that these institutional efforts have not been accompanied by parallel trends in residential segregation. The analysis by Taeuber and Taeuber (1965) and its extension (Sorenson, Taeuber and Hollingsworth 1975) report only a modest decrease in the average value of dissimilarity index calculated for 109 central cities between 1960 and 1970. According to Schnare, who analyzed neighbourhood segregation for 130 Standard Metropolitan Statistical Areas, the tendency toward segregation of blacks and whites increased over the 1960's (Schnare 1978). Similar results have been obtained by another study which examined segregation in 24 metropolitan areas (de Leeuw, Schnare and Struyk 1976). This tendency of increasing segregation is likely to compound the disadvantages of blacks by limiting their housing, economic and educational opportunities, and thus presents an important challenge to public policy.

Existing analyses of the causes of residential segregation, most of which relied on 1960 and pre-1960 data, present conflicting views and evidence. One of the widespread beliefs regarding residential segregation is that the physical separation of different races is voluntary and attributable to the differences in economic status between races. The poverty hypothesis, an explanation based on differential economic status between races, can be derived from both ecological and economic literature. The ecological approach relates the spatial pattern of population to social relations (Park 1967). Earlier works by Duncan and Duncan (1955) and Lieberson (1963) provide supports for this approach. More recently, Marshal and Jiobu, (1975) have conducted a path analysis of residential segregation for 149 urban places in 1960, and found the relative status of blacks to be a significant determinant of segregation. Similarly, the economic models of segregation developed by Bailey (1959, 1966), Becker (1957), and Muth (1969) suggest that segregation can be explained by differences in income or taste. One of the most rigorous analyses of the effect of economic factors on segregation, performed by Pascal (1967), reports that about 40 to 50 per cent of the residential segregation in Chicago and Detroit for 1950 and 1960 was explained by socio-economic variables.

The poverty hypothesis is basically derived from a market explanation of urban location. In this sense, the preference hypothesis shares common intellectual elements with the poverty hypothesis. The explanation based on the preference hypothesis suggests that segregation is a result of differences in consumer preferences between races. The models by Bailey (1959), Becker (1957) and Muth (1969) imply that the differential tastes can lead to segregation without white dominance in markets, discrimination and other causes of market failure. If whites have a greater tendency of aversion to residing in a mixed neighbourhood, they pay more for housing in white areas and create segregation.

Another explanation of voluntary segregation is the self segregation hypothesis. While the preference hypothesis postulates aversion among whites to racially mixed areas, the self-segregation hypothesis stresses the tendency among blacks to cluster together. But some believe that evidence for the self-segregation is not well-founded. Public opinion polls have shown consistently that a large proportion of blacks prefer integrated neighbourhoods (Brink and Harris 1969).

An opposing view of voluntary segregation emphasizes the role of discrimination as a prominent factor contributing to residential segregation. Discrimination is usually defined as actual implementation of prejudiced attitude, and can be practiced at individual as well as institutional levels. At individual levels, discrimination is behaviour of one person (or persons) of denying rights of others. An individual homeowner may refuse to sell his property to particular people based on his prejudice. At institutional levels, discrimination can be exercised by real estate intermediaries and financial institutions. It could take both explicit and subtle forms of discriminatory practices such as price premium for blacks, redlining and interest discrimination.

While earlier economic models by Bailey, Becker and Muth showed that it is possible to obtain racial segregation in the absence of market imperfection, more recent theoretical analyses suggest that no stable pattern of spatial distribution of blacks and whites can be obtained in the long run without discrimination (Yinger 1976, Courant and Yinger 1977). The importance of prejudice and discrimination in segregation has been recognized by various empirical analyses. In his study of segregation in Chicago and Detroit for 1950 and 1960 Pascal (1967) notes that attitudinal factors have significant effects on segregation. Taeuber calculates the expected proportion of black families in Chicago and Cleveland in 1960 using family income as a sole determinant of residential choice, and concludes that the overall effect of economic factors on residential segregation is tenuous (Taeuber 1965, 1968). Tomlinson (1969) also maintains that economic and educational gains made by blacks did not have much impact on altering the pattern of segregation. Similarly, Kain and Quigley (1975) after a careful review of existing evidence conclude that racial prejudice and discrimination are the major causes of racial residential segregation.

Substantial research efforts have been focused on the housing price differentials between blacks and whites to measure the extent of discrimination in housing markets. A number of studies using different methods of analysis provide empirical evidence for a price markup for blacks (Rapkin 1966, Ridker and Henning 1967, Muth 1969, King and Miesykowski 1973, Quigley 1974, Straszheim 1974). Only a few have implied different results (Bailey 1966, Lapham 1971). Researchers who subscribe to the discrimination hypothesis believe that the segregation is achieved by price discrimination which prevents blacks from entering white neighbourhoods. However, those who reject the discrimination hypothesis relate the existence of a price markup to other factors. For example, Muth (1969) suggests the price difference is due to the differences in operating costs. Muth (1974) further argues that due to the variety of characteristics of dwelling units and historical nature of black-occupied housing in American cities, it is difficult to obtain reliable evidence on price differentials between races from family housing expenditure data.

EMPIRICAL ANALYSIS

A Model and Data

Much of the controversy regarding the causes of residential segregation is centred around the relative importance of economic forces and attitudinal factors. Therefore, the primary objective of an empirical investigation should be the evaluation of the relative significance of the two determinants.

For the poverty hypothesis to be correct no segregation should be observed for a city with an identical pattern of income distribution for blacks and whites. If economic forces are the key determinants of residential segregation, the intercity variations in segregation should accrue mainly to the differences in the relative economic status of blacks among cities. On the other hand, if attitudinal factors such as prejudice and discrimination are primarily responsible for the segregation by race, indicators of these factors would explain most of the intercity variations in segregation. Therefore, a simple form of regression model explaining intercity variations in segregation can be formulated by using a measure of segregation as dependent variable and indicators of economic forces and attitudinal factors as explanatory variables.

Two additional considerations should be made in formulating the model. First, it is necessary to include variables describing housing market conditions since housing markets may play a rôle in defining constraints on black housing opportunities. The inclusion of housing market variables would also enable us to observe the delicate relationship between housing opportunities and segregation. The other consideration is the role of demographic factors in residential segregation. Theoretically, this consideration is based on the observation that differential demographic conditions may result in different levels of heterogeneity and attitude. First, it has been suggested that the level of internal differentiation of cities increases with city size (Kish 1954, Betz 1972). Therefore, it can be hypothesized that city size has systematic association with the level of segregation. Second, as some sociological studies suggest, increases in the relative size of black population may be regarded as a threat to opportunities for whites and as a result create stronger discriminatory behaviour (Blalock, 1956). Third, it is also possible that the relative growth rate of the black population exercises a systematic influence on white attitudes. Marshall and Jiobu (1975), for example, have postulated that a higher black-white growth differential would lead to higher levels of segregation due to the strengthening of discriminatory behaviour among whites as a response to an influx of blacks.

To summarize the above discussion, a regression model investigating factors affecting the intercity variation in segregation takes a general form of the following:

$$SEG = f\ (ECON,\ ATTI,\ HOUS,\ DEMO),$$

where SEG is an indicator of segregation, ECON is the relative economic status of blacks, ATTI is attitudinal variables, HOUS is housing market variables and DEMO is variables describing demographic conditions.

For the dependent variable, two measures have been chosen. One is the dissimilarity index between blacks and whites in 1970 (BWSEG) and the other is the dissimilarity index between non whites and whites in 1970 (NWSEG).

Two alternative indicators are selected to represent the relative economic status of blacks. RBISM is measured in terms of black family median income as per cent of white family median income in a SMSA. RBICC is black family median income as per cent of white family median income in a central city. Both variables enter regression equations in logarithmic forms.

It is extremely difficult to obtain variables accurately describing attitudinal factors. Currently, there are no quantitative measures of prejudice or discriminatory behaviour available for cities. As a proxy, a set of regional dummy variables are created. Cities under consideration are divided by four regions of the U.S. Taking the South as a reference, the dummy variable EAST takes the value of one if a city belongs to the Northeastern States. WEST is one if a city belongs to the West. If a city belongs to the North Central States, the regional dummy variable NCEN takes the value of one. These regional dummy variables are intended to capture the effect of various factors including attitudinal differences among regions not explained by other variables in the model. Although the regional dummy variables are far from being accurate measures of attitudes, a number of existing studies suggest that there are noticeable differences in attitudes among regions (Marx 1969, Selznick and Steinberg 1969). Since the effects of other determinants which may vary among regions are controlled in our models, the regional dummy variables seem to be a reasonable representation of attitudinal factors.

For housing market conditions, the variable RENT has been created. RENT is the average rent for a central city. It is calculated as an wieghted average of median contract rent for renters and imputed value of rent for homeowners.

Four demographic variables have been selected. POPUL is the logarithmic transformation of the population size of a city. PERBL is the percent black population in a city. PERBL is used for regression models using the segregation index between blacks and whites (BWSEG) as dependent variables. For models using the segregation index between nonwhites and whites (NWSEG) as dependent variable, the per cent nonwhite population (PERNW) is used alternatively. Finally, an indicator of relative growth rate of blacks is employed. GRODI is the logarithmic value of nonwhite net migration less white net migration divided by the total city population. Table 11.1 presents the list of variables and their definitions, and Table 11.2 shows the summary statistics of the variables.

Data for this analysis are collected from three different sources. The segregation indexes are obtained from Sorenson, Taeuber and Hollingsworth (1974). Data for GRODI are computed from Ritchey and Bishop (1974). All other data come from the 1970 Census of Population and Housing. Initially, data for central cities of fifty largest SMSA's were collected. But availability of data reduced the number of cities to forty-three.

Results

Using the framework described above, ordinary least square regressions having various functional forms have been estimated. The final set of regressions reported here takes a semi-log form with independent variables with or without logarithmic transformations. The reasons for the choice of this particular form are to avoid multicollinearity problems and to obtain a better goodness-of-fit. Table 11.3 presents regression results for factors affecting segregation between blacks and whites and Table 11.4 reports those for nonwhites and whites.

Table 11.3 shows four regression models using the segregation index between blacks and whites (BWSEG) as dependent variable. The models I and II include only the economic variables and the regional dummy variables. The two variables describing the relative economic status of blacks are never significant in the two models. However, the regional dummy variables are consistently significant. EAST and WEST are statistically significant at least at the one per cent level. Since the South is taken as a reference the negative signs of EAST and WEST suggest that, other things being equal, cities in the Northeastern States and the West are less segregated. The variable NCEN is less significant in comparison to EAST and WEST. The models I and II suggest that economic factors count little while regional differences have strong effects on segregation. But, the regional dummy variables as used in the models I and II cannot be interpreted as direct representation of differential attitudes among regions, since other variables not included in the equation may vary with regions.

TABLE 11.1

Variable Definitions

Dependent Variables	BWSEG	Taeuber Index between blacks and whites in 1970
	NWSEG	Taeuber Index between nonwhites and whites in 1970
Independent Variables	RBISM	Relative economic status of blacks in a SMSA in 1970. (Black family median income in a SMSA - White family median income in a SMSA) x 100% in Log.
	RBICC	Relative economic status of blacks in a city in 1970. (Black family median income in a central city - white family median income in a central city) x 100% in Log.
	EAST	A regional dummy variable. 1 if a city is in the Northeastern States.
	WEST	A regional dummy variable. 1 if a city is in the West.
	NCEN	A regional dummy variable. 1 if a city is in the North Central States.
	RENT	Log of the average rent for a central city in 1970. (proportion of renters x median rent) + (.12 x proportion of owners x median housing price in Log.
	POPUL	Log of the population of a central city in 1970.
	PERBL	% black population in a central city in 1970.
	PERNW	% nonwhite population in a central city in 1970.
	GRODI	An indicator of relative growth rate of nonwhites. (Nonwhite net migration - White net migration) - population of the central city in 1960.

TABLE 11.2

Summary Statistics for Variables

Variable	Mean	Standard Deviation
BWSEG	86.38	6.42
NWSEG	81.65	9.28
RBISM	4.14	.10
RBICC	4.21	.12
EAST	.21	.41
WEST	.16	.37
NCEN	.28	.45
RENT	7.10	.18
POPUL	13.2	.79
PERBL	24.30	14.23
PERNW	25.84	13.60
GRODI	.09	.08

To control the effects of other variables housing and demographic variables are included in models III and IV. The role of the economic variables (RBISM and RBICC) remain insignificant. Slight changes occur with the regional dummy variables when housing and demographic variables are added. NCEN which is significant in the first two models is no longer significant. However, the other two regional dummy variables, EAST and WEST remain statistically significant and show consistent results in the signs and magnitudes of their coefficients. The significance of the two regional dummy variables, EAST and WEST, is consistent with implications of the previous studies on prejudice. The studies by Marx (1969) and Selznick and Steinberg (1969) have shown that attitudinal conflicts between races are greater in the South than in the North.

Among the housing and demographic variables only the average rent (RENT) shows statistically significant results. None of the demographic variables turn out to be statistically meaningful. As noted in the previous section, evidence regarding the relationship between housing prices and segregation is conflicting. According to the preference hypothesis, prices are lower in inner black areas than boundary and white areas in the long run. In contrast, the discrimination explanation suggests that blacks pay more for housing due to barriers to enter white submarkets. The two opposing approaches, however, agree that prices rise in the short run temporary disequilibrium. An explanation of this phenomenon is that in the process of transitional integration blacks outbid whites which leads to higher prices. The regression results for RENT seem to confirm this position. The constantly negative sign of RENT suggests that residential integration is associated with higher housing prices in the short run.

TABLE 11.3

Regression Models for Segregation between Blacks and Whites
(Dependent variable = BWSEG)

	I	II	III	IV
Constant	78.00** (1.74)	82.99** (2.19)	178.53*** (2.62)	149.55** (2.40)
RBISM	3.20 (.29)		-5.48 (-.51)	
RBICC		1.95 (.21)		.19 (.02)
EAST	-10.75*** (-4.73)	-10.85*** (-4.20)	-8.52*** (-3.17)	-9.00*** (-3.10)
WEST	-8.68*** (-3.02)	-8.48*** (-3.14)	-5.88* (-1.59)	-6.86** (-1.93)
NCEN	-4.21** (-1.69)	-4.17* (-1.54)	-1.22 (-.43)	-2.14 (-.71)
RENT			-11.10** (-2.10)	-10.39** (-1.97)
POPUL			1.14 (1.13)	1.21 (1.19)
PERBL			-.07 (-1.03)	-.07 (-.97)
GRODI			-10.50 (-1.01)	-10.06 (-0.96)
R^2	.43	.42	.56	.56

t- statistics are in parentheses
*** significant at the 0.01 level
** significant at the 0.05 level
* significant at the 0.10 level

TABLE 11.4

Regression Models for Segregation Between Non-whites and Whites
(Dependent variable = NWSEG)

	I	II	III	IV
Constant	85.51*	85.26**	189.48**	160.80**
	(1.57)	(1.85)	(2.16)	(2.01)
RBISM	.66		-5.64	
	(.05)		(-.40)	
RBICC		.71		.00
		(.06)		(.01)
EAST	-10.69***	-10.76***	-6.66**	-7.12**
	(-3.87)	(-3.42)	(-1.92)	(-1.90)
WEST	-20.04***	-20.05***	-14.58***	-15.58***
	(-5.73)	(-6.10)	(-3.15)	(-3.53)
NCEN	-3.84	-3.89	.28	-.63
	(-1.27)	(-1.18)	(.07)	(.16)
RENT			-12.53**	-11.85**
			(-1.84)	(-1.74)
POPUL			.62	.69
			(.48)	(.52)
PERNW			.05	.06
			(.57)	(.62)
GRODI			-15.60	-15.13
			(-1.16)	(-1.12)
R^2	.59	.59	.65	.65

t- statistics are in parentheses
*** significant at the 0.01 level
** significant at the 0.05 level
* significant at the 0.10 level

In the face of studies indicating the importance of demographic variables, the present result showing insignificance of demographic variables is rather surprising. However, it should be noted that the previous studies which analyzed the effect of demographic variables on segregation (for example, Marshall and Jiobu 1975) did not control the effect of regional differences in attitudes directly. In our experiments with various regression equations some of the demographic variables were indeed significant when used without the regional dummy variables. But when the regional dummy variables were added, the demographic variables lost their statistical significance.

Fairly similar results have been obtained for the regression models explaining intercity variation in the segregation between nonwhites and whites. (Table 11.4) The dependent variable in the model is the dissimilarity index between nonwhites and whites. The only modification in the independent variables is the use of percent nonwhite population (PERNW) in place of percent black population (PERBL).

None of the two economic variables (RBISM and RBICC) measuring the relative economic status of the black population are significant. This result, along with the result for black-white segregation, serves to disapprove the poverty hypothesis.

Consistently significant variables are the two regional dummy variables (EAST and WEST) and the housing variable (RENT). Both EAST and WEST are significant at least at the 5% level throughout all models. Although the coefficients for the two regional dummy variables consistently show a negative sign for both black-white and nonwhite-white segregation, the differences in their magnitude deserve some discussion. Based on the models III and IV the average effects of EAST and WEST are -8.76 and -6.37 for black-white segregation respectively. For nonwhite-white segregation the values are -6.89 for EAST and -15.08 for WEST. The effect of EAST is slightly stronger for black-white segregation and the effect of WEST is substantially stronger for nonwhite-white segregation. These differences may reflect the different history of race relations among regions. Historically, the regions of the United States have shown markedly different patterns of racial relations (Jones 1972). On the one hand, the Northeastern States have relatively longer exposure to black immigration than the West. On the other hand, cities in the West have had longest history of migration of nonwhites other than blacks. Two alternative explanations may be offered to account for the relationship between historical differences in racial relations and segregation. First, it is possible that the different historical context of migration has resulted in differential attitudes among regions. For example, the longer history of the Western cities in having contact with nonwhites other than blacks may have resulted in relatively lower levels of prejudice and discriminatory behaviour on the part of whites. Alternatively, nonwhites in the West, due to the longer history of immigration, may have a better ability in dealing with discrimination against them.

The estimated result for RENT for nonwhite-white segregation shares little difference with that for black-white segregation. Finally, no statistically meaningful result is obtained for demographic variables.

SUMMARY AND IMPLICATIONS

The results of the empirical analysis presented in this paper can be summarized as follows: First, the analysis suggests that economic factors contribute little to the explanation of intercity variation in the level of racial segregation. Second, the regional differences in attitudes and other factors not explained by economic and demographic variables appear to have significant impacts on residential segregation. Third, while it is difficult to establish a clear casual relationship between segregation and housing market conditions, the result implies that residential integration has positive association with housing prices in the short run. Fourth, demographic factors such as population size, the relative size of the black population, and the relative growth rate of nonwhites do not seem to have substantial effects on segregation. Fifth, differences in the effect of the regional variables between black-white and nonwhite-white segregation are observed. It appears that the differences are due to the different history of race relations among regions.

Overall, the present study provides little evidence to support the poverty hypothesis. The attitudinal factors and housing market conditions seem to be the most important factors determining variations in both black-white and nonwhite-white segregation. This result is in line with the conclusion by Taeuber (1965, 1968), Tomlinson (1969), and Kain and Quigley (1975) who attribute segregation mainly to discrimination.

Probably one of the most important policy implications that can be drawn from the analysis is the need for localization of public policies dealing with segregation. The marked regional differences in segregation which may be caused by differential attitudes, institutional systems, and historical factors, cannot be effectively dealt with by a uniform program designed at the federal level. It seems the greater variations in housing and community development policies should be allowed to effectively cope with the problem of segregation.

This study also suggests that the improvement in the economic status of blacks and nonwhites is not likely to be accompanied by a reduction in segregation. This implies that more public policy efforts should be directed to anti-discrimination measures, insofar as the reduction in segregation remains a principle policy objective.

REFERENCES

Bailey, M. Note on Economics of Residential Zoning and Urban Renewal, **Land Economics**, 35 (Aug. 1959), 288-90.

Bailey, Effects of Race and Other Demographic Factors on the Values of Single-Family Homes, **Land Economics**, 42 (May 1966), 215-22.

Becker, G. **The Economics of Discrimination**, Chicago: University of Chicago Press (1957).

Betz, D.M. The City as a System Generating Income Equality, **Social Forces**, 51 (Dec. 1972), 192-8.

Blalock, H.M. Economic Discrimination and Negro Increase, **American Sociological Review**, 21 (Oct. 1956), 284-8.

Brink W. and Harris, L., Breaking the Vicious Circle, in **Race and Poverty: The Economics of Discrimination**, ed. John Kain Englewood Cliffs, N.J.: Prentice Hall, (1969), pp. 139-45.

Courant P. and Yinger, J., On Models of Racial Prejudice and Urban Residential Structure, **Journal of Urban Economics**, 4 (Apr. 1977), 272-91.

Daniels, C.B., The Influence of Racial Segregation on Housing Prices, **Journal of Urban Economics**, 2 (Jan. 1975), 105-22.

De Leeuw, F., Schnare A.B., and Struyk, R.J., Housing, in **The Urban Predicament**, eds. W. Gorham and N. Glazer Washington, D.C.: The Urban Institute, (1976).

Duncan O. and Duncan, B., Residential Distribution and Occupational Stratification, **American Journal of Sociology**, 60 (Mar. 1955), 493-503.

Haugen R., and Heins, A., A Market Separation Theory of Rent Differentials in Metropolitan Areas, **Quarterly Journal of Economics**, 83 (Nov. 1969), 660-72.

Jones, J.M., **Prejudice and Racism** Reading, Mass.: Addison-Wesley, (1972).

Kain J., and Quigley, J., **Housing Markets and Racial Discrimination: A Microeconomic Analysis**, (New York: National Bureau of Economic Research (1975).

King T. and Mieszkowski, P., Racial Discrimination, Segregation, and the Price of Housing, **Journal of Political Economy**, 81 (May-Jun. 1973), 590-606.

Kish, L., Differentiation in Metropolitan Areas, **Amer. Soc. Rev.**, 19 (Aug. 1954), 388-98.

Lapham, V., Do Blacks Pay More for Housing? **Journal of Politi-**

cal Economy, 79 (Nov.-Dec. 1971), 1244-57.

Lieberson, S., **Ethnic Patterns in American Cities,** New York: Free Press, (1963).

Marshall H., and Jiobu, R., Residential Segregation in U.S. Cities: A Causal Analysis, **Social Forces,** 53 (Mar. 1975), 449-60.

Marx, G.T., **Protest and Prejudice: A Study of Belief in the Black Community** New York: Harper and Row (1969).

Morrill, R., The Negro Ghetto: Problems and Alternatives, **Geographical Review,** 55 (Jul. 1965), 339-61.

Muth, R., **Cities and Housing: The Spatial Pattern of Urban Residential Land Use,** Chicago: University of Chicago Press (1969).

Muth, R.F., Residential Segregation and Discrimination, in **Patterns of Racial Discrimination,** Vol. 1, **Housing,** eds. von Furstenberg, G.B., Harrison B., and Horowitz A., (Lexington, Mass: Lexington Books, 1974).

Park, R., The Urban Community as a Spatial Pattern and Moral Order, in **Robert E. Park: On Social Control and Collective Behavior,** ed. Ralph Turner Chicago: University of Chicago Press (1967).

Pascal, A., The Economics of Housing Segregation, **Rand Research Memorandum No. RM-5510**-RD Santa Monica, Calif.: Rand Corporation, (1967).

Quigley, J.M., Racial Discrimination and the Housing Consumption of Black Households, in von Furstenberg, Harrison and Horowitz.

Rapkin, C.,Price Discrimination against Negroes in Rental Housing Market, in **Essays in Urban Land Economics,** ed. Kain J., Los Angeles: University of California Press, (1966).

Ridker R., and Henning, J., The Determinants of Residential Property Value with Special Reference to Air Pollution, **Review of Economics and Statistics,** 44 (May 1967), 246-55.

Ritchey P.N., and Bishop, B.L., **Components of Change for the Adult Populations of Cities by Age, Sex and Color** Oak Ridge, Tenn.: Oak Ridge National Laboratory (1974).

Rose-Ackerman, S., Racism and Urban Structure, **Journal of Urban Economics,** 2 (Jan. 1975), 85-103.

Schelling, T., Dynamic Models of Segregation, **The Journal of Mathematical Sociology,** 1 (Jul. 1971), 143-86.

Schnore, L., **Class and Race in Cities and Suburbs** Chicago: Markham, (1972).

Schnare, A.B., **The Persistence of Racial Segregation in Housing,** Washington, D.C.: The Urban Institute, (1978).

Selznick G.J., and Steinberg, S., **The Tenacity of Prejudice: Anti-Semitism in Contemporary America,** New York: Harper and Row, (1969).

Sorensen, A., Taeuber K.E., and Hollingworth Jr., L., Indexes of

Racial Residential Segregation for 109 Cities in the United States, 1940-1970, **Sociological Focus,** 8 (Apr. 1975), 125-42.

Straszheim, M., Racial Discrimination in the Urban Housing Market and Its Effects on Black Housing Consumption, in von Furstenberg, Harrison and Horowitz.

Taeuber, K.E., Residential Segregation, **Scientific American,** 213 (Aug. 1965), 12-19.

Taeuber, K., The Effect of Income Redistribution on Racial Residential Segregation, **Urban Affairs Quarterly,** 4 (Sep. 1968), 5-14.

Taeuber K., and Taeuber, A., **Negroes in Cities,** Chicago: Aldine, (1965).

Tomlinson, R., **Urban Structure: The Social and Spatial Character of Cities,** New York: Random House, (1969).

Van Valey T.L., and Roof, W.C.,Measuring Residential Segregation in American Cities: Problems of Intercity Comparison, **Urban Affairs Quarterly,** 11 (Jun. 1976), 453-68.

Yinger, J., Racial Prejudice and Racial Residential Segregation in an Urban Model, **Journal of Urban Economics,** 3 (Oct 1976), 383-96.

TWELVE

E van Loggerenberg and JN Scheepers

Unique Residential Areas in Metropolitan Witwatersrand

Towards Interpreting Multi-dimensional Outcomes

THE PROBLEM: DIFFERENTIATED RESIDENTIAL PATTERNS IN URBAN SOCITIES

Since human society is an increasingly urban phenomenon (Timms, 1971, 253), human behaviour finds it dominant expression in the city. However, if the spatial expression of human behaviour is to be understood and applied to practical planning and management, the complexities of urban societies should be monitored by professional geographers with their bias towards these problems. Throughout the world, increasing urban populations give rise to undifferentiated settlement patterns with a pot-pourri of buildings and groups of people. Without an understanding of the underlying principles, effective improvements in urban environments and the provision of community needs cannot be achieved.

The urban population of South Africa as a percentage of its total population has grown from 26 per cent in 1911 to almost 48 per cent in 1980. The average growth of metropolitan areas in South Africa is presently 5.3 per cent per annum—much higher than the metropolitan areas of the developed world (Northam, 1979, 72). This fact is explained by South Africa's position as part of the developing world. Here urbanization started at a later stage and continues to grow at an increasing rate, as opposed to developed countries which have stabilized at an advanced stage of urbanism.

Against this background of high growth rates the question arises as to whether the underlying spatial structure can still be reliably characterized to an acceptable level.

Many urban Geographers consider their most worthwhile

goal to be the application of their results towards the future planning of the city (Garner, 1966, 210). If human interaction in urban space leads to the creation of specific residential patterns, then such patterns can be identified and described objectively.

This research aims towards:

(i) The identification of dimensions (from a multivariate set of data) which are fundamental to the important characteristics of the population. This means that the actual statistical variation should be technique-independent to explain the real world situation.

(ii) The identification of spatial patterns that result from (i) above. Problems in describing unique factors or dimensions disclosed by multivariate analysis are well-known to researchers in this field.[1] This study illustrates that misinterpretation of analytical results may lead to conclusions that cannot withstand the test of reality.

THE AREA: A MULTI-NODAL METROPOLIS

The Witwatersrand is South Africa's largest metropolitan area and in 1970 accounted for 29 per cent of the gross domestic product, and for 29 per cent of the White urban population. The region's urban population has grown from a total of 1,708,000 in 1951 to 2,629,000 in 1970, of whom at the latter date 942,000 (35.8 per cent) were White. The White population growth rate increased from 1.7 per cent per annum between 1951 and 1960 to 2.4 per cent per annum between 1960 and 1970, and was obviously strongly influenced by the high rate of immigration to South Africa during those years (Johannesburg, 1972). As a result of these trends, and strong economic growth coupled with income growth, the expansion of land in White residential townships increased annually by 3.2 per cent compared with a lower rate of White population increase (Fair, 1978, 3). Consequently, as the suburbs sprawled outwards and motor cars and new highways enabled people to live in more remote surroundings in their search for open space, the overall density of the White population decreased.

Physically, the Witwatersrand extends in an east-west linear pattern for approximately 100 kilometres, while its north-south extension does not exceed 35 kilometres. Johannesburg forms the region's dominant centre and is surrounded by 17 other local authorities contiguous to one another so that the Witwatersrand on the whole discloses a multi-nodal pattern.

THE ANALYSIS

1. The data matrix

In order to achieve both aims, data which contains the characteristics of the population for the extraction of broader dimensions which underlie these characteristics, is indispensable. Socio-economic data is normally suitable for such analyses. The study of income, class and status as spatial variables has been pursued by social geographers and is linked closely to structural theories of city patterns (Carter, 1975, 249). Therefore, the 1970-census data was used as a starting point. Data of the White, Brown and Indian population groups[2] of the Witwatersrand containing socio-economic variables pertaining to individuals, families and dwelling units were then used, and mainly statistically reduced to 30 variables in 532 suburbs (van Loggerenberg, 1982B):

1. Age of head of family
2. Size of family
3. Female family members
4. Age: above 65 years
5. Age: below 15 years
6. Home language: Afrikanns
7. Home language: English
8. Home language: Other
9. Religion: Roman Catholic
10. Religion: Hebrew
11. Religion: Afrikaans denominations
12. Place of birth: Europe
13. Level of formal education: lower than Std. 10
14. Level of formal education: post high school
15. Number of families per dwelling unit
16. Type of dwelling: flats
17. Occupation: one dwelling
18. Number of bedrooms per dwelling unit
19. Number of bathrooms per dwelling unit
20. Number of outside rooms for servants
21. Number of full-time servants
22. People in professional occupations
23. People in sales occupations
24. People in production and transport occupations
25. People in administrative and clerical occupations
26. People in manufacturing occupations
27. People in mining occupations
28. People in commercial and financial occupations

29. People in a low income category (R 2,000 p.a.)
30. People in a high income category (R 6,000 p.a.)

2. Dimensional analysis

A statistical approach was employed with regard to the analyses of the basic characteristics of the data. Five factor analytic models, viz., principal components analysis, factor analysis, maximum likelihood analysis (all with a correlation matrix as basic input), principal components and factor analysis (both with a covariance matrix as input) were employed to analyse the data. After this the results were rotated to the orthogonal and oblique principles. Correlative analysis was employed to ascertain the resemblance among the components and factors extracted. Canonical correlation analysis was used to establish the correspondence of dimensions extracted from the factor analytic models. The following seven dimensions were then identified as underlying the real characteristics of the data and were considered as not technique-dependent (van Loggerenberg, 1982A):

Dimension 1: Older immigrants particularly those in business and professional occupations.
Dimension 2: Afrikaans working class in the child-bearing phase.
Dimension 3: High-class single-unit residence.
Dimension 4: Large families characterized by female and older family members.
Dimension 5: Rented housing for mine workers in particular.
Dimension 6: Flat residence by older people.
Dimension 7: Low-class residence and the working class.

Together, these seven dimensions accounted for 83.1 per cent of the variation with regard to an input using a correlation matrix, and 97.5 per cent when using a covariance matrix as input.

3. Spatial Analysis

To evaluate the residential patterns in the area, each of the seven dimensions was mapped using the SYMAP, Version 5.20 computer programme. The factor scores which were used to classify the 532 different suburbs, consisted of standardised values.

In order to establish classes for each map, class intervals had to be chosen so that the most significant maps possible could be produced. The starting point may be the frequency of values around the average or mode (Scheepers, 1978, 95 and Evans, 1977, 101), arbitrary cutting points which may vary from map to map (van der Merwe and Zietsman, 1977, 2), or cutting points resulting in more or less the same number of suburbs in every category (Hart, 1975, 52). When frequency diagrams are constructed using the standardised values of the suburbs, skew distributions result (See Figure 12.1), and since the skewness of the distribution varies, standard cutting points for all sets of scores would seem to have disadvantages and different cutting points can be used (Robson, 1969, 169). The important issue, however, is the aim of a specific classification.

In this case the major aim was to identify those suburbs in which a specific dimension dominated (those with high positive factor scores), those in which the opposite prevailed (those with high negative factor scores), and those suburbs in which no association existed with a particular dimension (those with factor score values around nil). Consequently, the decision was made that three classes would suffice to contain high positive outliers, high negative outliers and those which could be considered as the most frequent "normal" or "modal" suburbs. To identify these three classes, the frequency diagrams (Figure 12.1) were carefully studied. The positive and negative outliers were defined as those points on the frequency diagram where distinct knick points would appear, if a continuous curve were constructed over the vertical lines.

These knick points were indicated on the diagrams by means of arrows which were considered to be the various limits within which the most frequent "modal" phenomena is found, with outliers on either side thereof. As a consequence, different cutting points as indicated on each frequency for each dimension were used so that the pattern of a specific dimension clearly emerged. Positive factor scores are indications of the incidence of variables which have high loadings on specific dimensions, and vice versa. This boils down to the fact that positive scores indicate that variables with negative loadings do not associate with those suburbs. The opposite holds for the negative scores. The occurrence of positive and negative facets of dimensions could thus be identified in space.

Figures 12.2 (a) to (g) indicate the spatial variation of each dimension. From these maps it is clear that only a few suburbs qualify as indicative of a significant uniqueness in relation to that specific dimension which is portrayed. Two conspicuous phenomena, however, need to be considered.

12.1 : FREQUENCY DIAGRAMS OF FACTOR SCORES

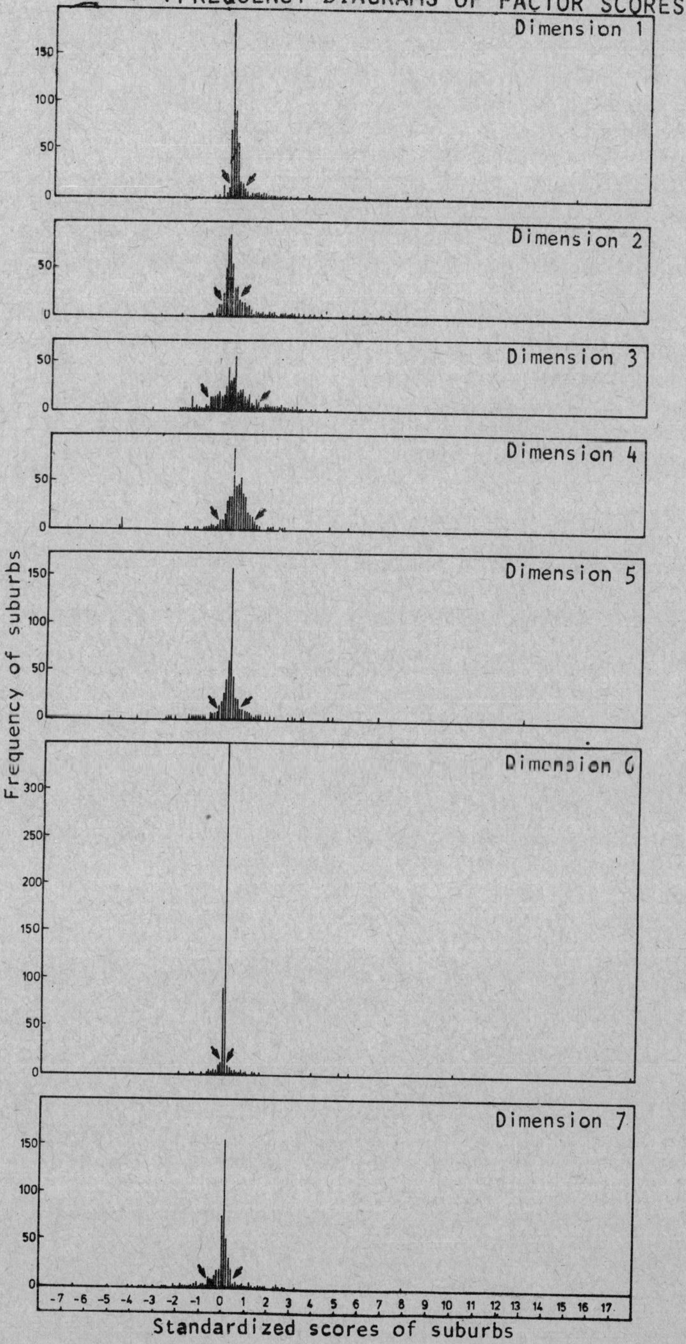

FIG. 12.2A : DIMENSION 1

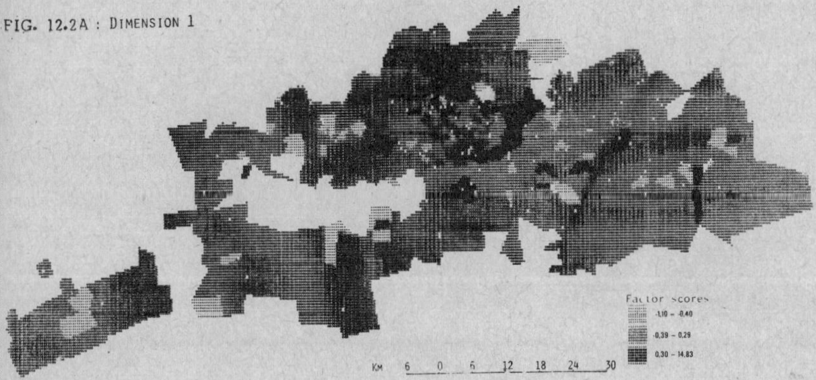

FIG. 12.2B : DIMENSION 2

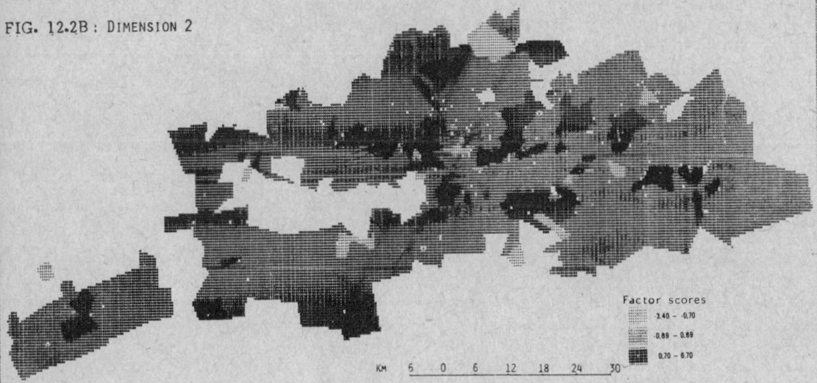

FIG. 12.2C : DIMENSION 3

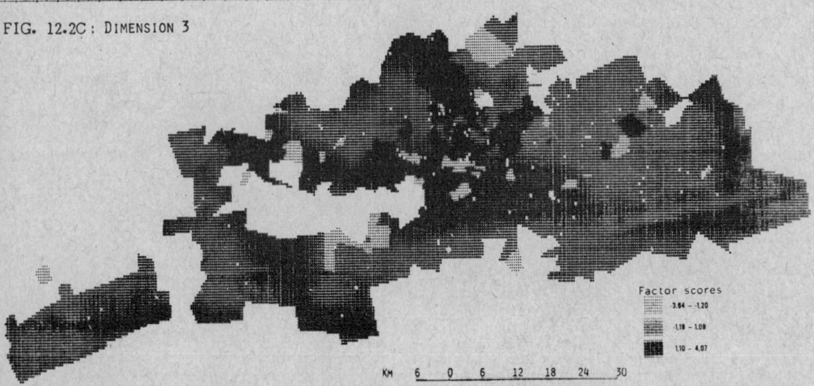

FIG. 12.2D: DIMENSION 4

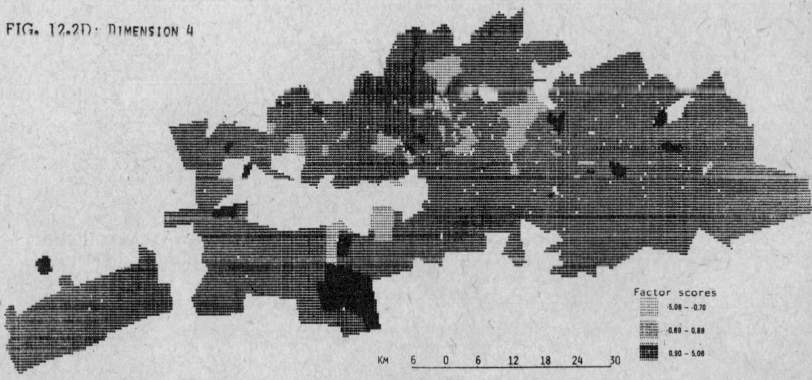

FIG. 12.2E : DIMENSION 5

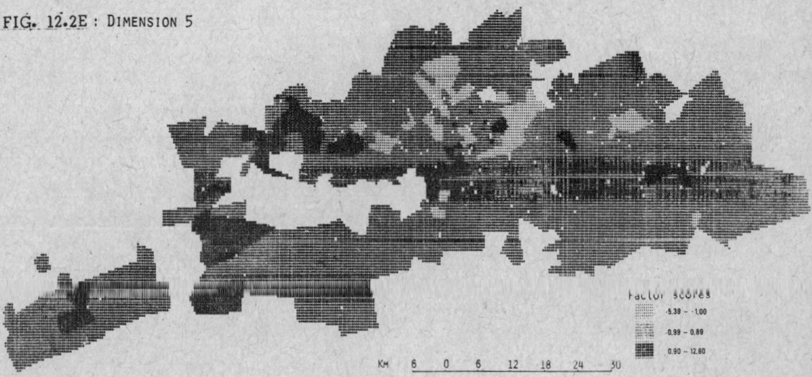

FIG. 12.2F: DIMENSION 6

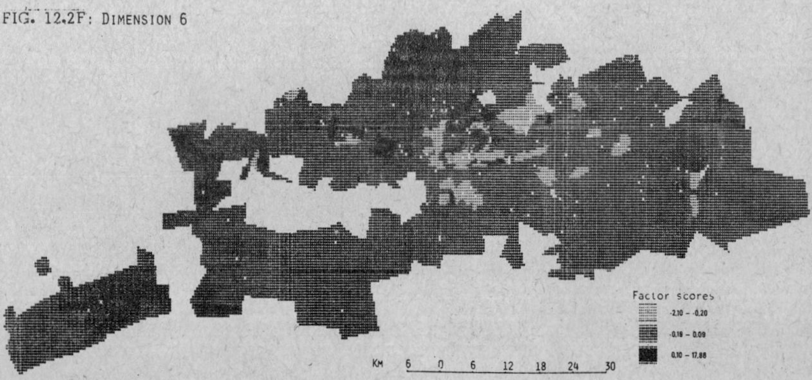

FIG. 12.2G: DIMENSION 7

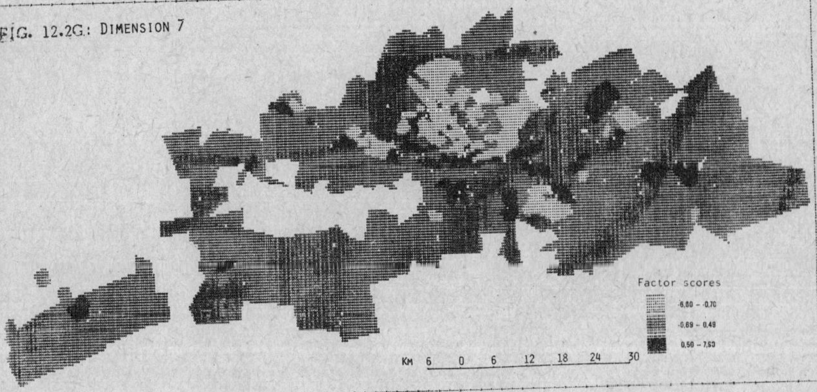

The first phenomenon is not a rare occurrence, because of the contents of the different dimensions. It is logical that some overlapping should exist between, e.g., dimension 2: Afrikaans working class in the child-bearing phase; and dimension 7: low-class residence and the working class. Some overlapping can also be expected to occur when the spatial structure of dimension 1: Older immigrants particularly those in business and professional occupations; and dimension 6: Flat residence by older people, is studied. Consequently, the detailed study of such occurrences can lead to the identification of suburbs with multicharacteristics, some of which will differ from one another according to the various combinations of dimensions found.

The second phenomenon is of great significance, especially when figure 12.3 is studied. Figure 12.3 indicates a combination of all the suburbs of all seven dimensions which have positive outliers. All the suburbs, whether only one dimension is present or whether overlapping dimensions occur, were mapped onto one figure. This resulted in the interesting fact that a considerable number of suburbs are not characterised by any dominant dimension. Such areas cannot be typified and can be regarded as non-unique.

CONCLUSION

The Identification of Non-unique Areas

The results of overlapping dimensions in certain suburbs, and the absence of unique characteristics, have implications for the theory of city structure as well as for planning management purposes. In the past, too many researchers were far too

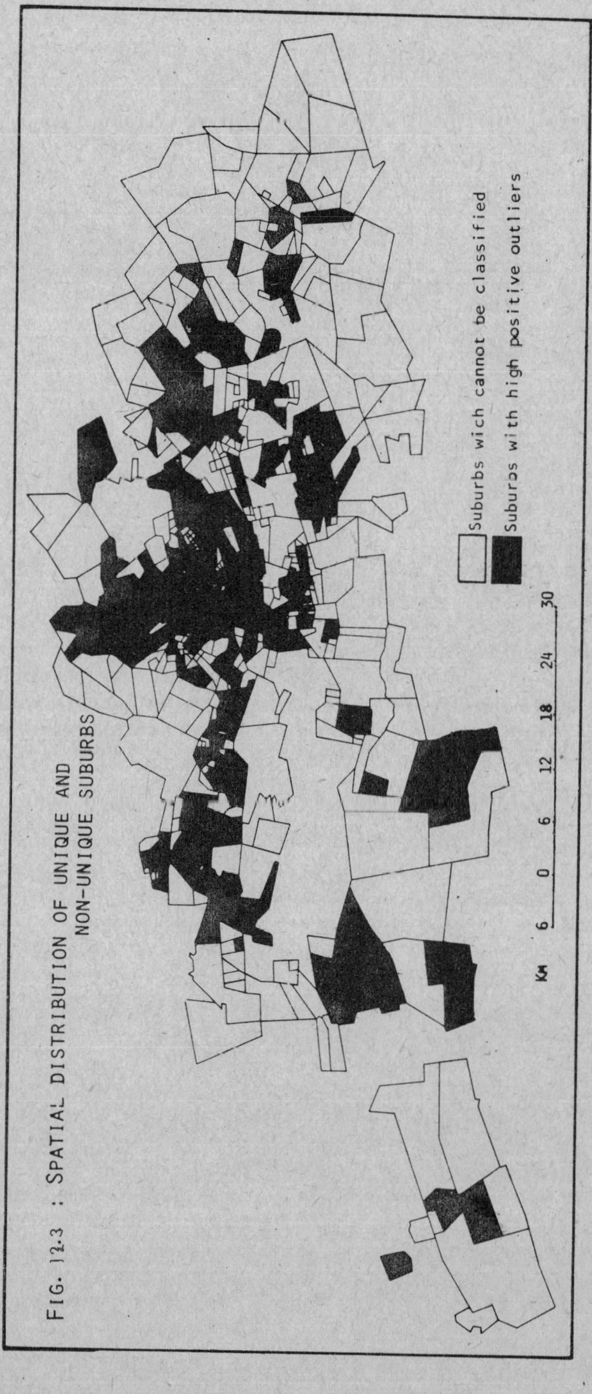

FIG. 12.3 : SPATIAL DISTRIBUTION OF UNIQUE AND NON-UNIQUE SUBURBS

readily inclined to fix a definite dimension or label to each and every suburb or zone in a city. However, the existence of non-unique, characterless, non-classifiable areas should be accepted as reality and the theories of city structure should be rephrased accordingly. The lament of Robson (1975, 4) that cities have changed in response to changes in technology and in the social structure of the population, but that urban theory has not kept abreast, is unfortunately still valid today.

FOOTNOTES

1. CF Berry, 1971; Robson, 1969; Rees,1970; Berry & Ress, 1969; Abu-Lughod, 1969; Hart, 1975; Borukhov et al., 1979; and many others.
2. Data on the Black population group was not available on the same basis.

REFERENCES

Abu-lughod, J. Testing the theory of social area analysis: The case of Cairo, Egypt. **American Sociological Review**, 34, (1969) 198-212.

Berry, B.J.L. Introduction: the logic and limitations of comparative factorial ecology. **Economic Geography**, 47 (2), (1971) 209-219.

Berry, B.J.L. and Horton, F.E. (eds.) **Geographic perspectives on urban systems.** Englewood Cliffs, Prentice-Hall (1970).

Berry, B.J.L. and Rees, P.H. The factorial ecology of Calcutta. **American Journal of Sociology**, 74, (1969), 445-491.

Borukhov, E. et al., The social ecology of Tel-Aviv - A study in factor analysis. **Urban Affairs Quarterly**, 15(2), (1979), 183-205.

Carter, H. **The study of urban geography.** 2nd edition. London. Edward Arnold (1975).

Evans, I.S. The selection of class intervals. **Transactions of the Institute of British Geographers.** (New Series), 2, (1977), 98-124.

Fair, T.J.D. **Spatial patterns and determinants of White residential land development on the Witwatersrand 1950-1975.** University of the Witwatersrand, Johannesburg (1978).

Garner, B.J. Aspects and trends of urban geography. **Journal of Geography**, 65, (1966) 206-211.

Hart, T. **The factorial ecology of Johannesburg.** University of the Witwatersrand, Johannesburg (1975).

Johannesburg **Urban Population Growth.** Report F.P. 40, City Engineers Department, Forward Planning Branch, Johannesburg Municipality (1972).

Mayer, H.M. and Kohn, C.F. **Readings in urban Geography.**

Chicago, University of Chicago Press (1959).

Northam, R.M. **Urban Geography.** 2nd edition. New York, John Wiley & Sons (1979).

Rees, P.H. Concepts of social space: Toward an urban social geography. (In Berry and Horton, 1970, 276-305).

Robson, B.T. **Urban analysis - a study of city structure.** London, Cambridge University Press (1969).

Robson, B.T. **Urban social areas.** London, Oxford University Press (1975).

Scheepers, J.N. Book review - Sociale Atlas van die Kaapstad Metropolitaanse gebied. **The South African Geographer,** 6(1), (1978) 93-95.

Timms, D.W.G. **The urban mosaic—towards a theory of residential differentiation.** London, Cambridge University Press (1971).

Van Der Merwe, I.J. and Zietsman, H.L. **Sosiale Atlas van die Kaapstad Metropolitaanse gebied** (1977).

Van Loggerenberg, E. **Residensiele patroonvorming aan die Witwatersrand: 'n meerveranderlike analise.** Rand Afrikaans University, C28, Johannesburg (1982A).

Van loggerenberg, E. Selection of variables for multi-variate analysis procedures. (To be published in **The South African Geographer**) (1982B).

THIRTEEN

Jean-Pierre Thouez and Peter M. Foggin

The Social Ecology of Cardiovascular Disease and Mortality

INTRODUCTION

The social geography of disease and mortality is to be viewed as the study of the diverse social factors, defined in the broadest sense, that influence the frequency of illnesses or deaths in time and space. Two phases are generally distinguished, primarily for pedagogical reasons, in the context of a geographical approach to this question. The first is that of determining the zones or areas of exceptionally high or low frequencies of a given type of morbidity or mortality and the thematic mapping of these zones either by hand or by computer assisted cartography. This exploratory phase usually succeeds in generating hypotheses as to possible risk factors that are related to the resulting spatial distributions. The second phase leads more particularly to the analysis of the relationships between the spatial distributions of morbidity or mortality and various postulated explanatory variables. For example, Max Sorre suggested, early in the development of what is now called medical geography that there exists a "pathogen complex" explaining the complex relationships between the presence of certain tropical diseases and the diverse environmental factors permitting their transmission (Sorre, 1951, p. 363). In a somewhat similar fashion Hunter (1980) relates the various factors involved through a modelling approach in order to specify action strategies for the prevention of onchoceriasis.

It is not surprising that the first attempts in medical geography had to do with tropical diseases, since the unusual spatial distributions they often manifested cried out for some kind of geographic analysis. Beyond this, many of these tropical illnesses could be traced to a fairly straighforward ecological chain process and so there were some direct links with the thinking of bio-geographers. More recently, geographers have become interested in various chronic diseases, characteristic of industrialized society, such as cardiovasculaire disease as well as the geographical aspects of the whole spectrum of neoplasmas (Pyle, 1971). The latter subject is sometimes referred to as geocancerology (Howe and Loraine, 1976) and is a rapidly developing field in itself. In all those areas of exploration and geographic research a multifactorial approach has to be maintained. This type of holistic interpretation was greatly developed by Jacques May (1966) and is still generally used by medical geographers (Meade, 1976). This approach views man and his environment in a complex network of relationships developed through his adaptative capacities. However, this human adaptation system is in no way deterministic since man acts as an agent of change in relation to his environment. Often these changes are far from being adaptive as can be seen in the case of the negative effects of industrial development. Transporting toxic products, for example, or even more clearly, environmental contamination or pollution create definite risks for exposed populations. Whereas the approach in the early attempts at medical geography was to define the geographic extent of a known pathogen complex, the present general approach to geographic research in this field involves the comparing of the geographical distribution of disease or mortality with various human activities and characteristics as they are projected over space. In a way this latter effort is very related to the initial approach of developing an understanding of the pathogen complex since pathogens, as a concept, were almost invariably related in some way to human activity.

In the present study, emphasis is placed on the social ecology of disease and mortality in the urban environment, leaving to one side physical factors even though the two realms are clearly related in time and space. The following presentation is three fold. First, a brief discussion of research bearing on social differences in relation to death statistics; second, an analysis of mortality and morbidity data for Montreal Island; and third, some methodological discussion in relation to the geographic study of illness and death.

1. Social inequality and death

The work of the Chicago-school sociologists such as Park and Burgess (1925) and somewhat more recently by geographers Harris and Ullman (1945) as well as other workers (Shevky and Bell, 1955, Berry and Kasarda, 1977) have shown that the social division of geographic space is not the product of pure chance but rather is the result of a certain number of fairly clearly defined parameters, such as socio-economic status and family status, for example. This has been shown particularly true for a good number of North American cities including Montreal (Foggin and Polese, 1976; Greer-Wootten, 1972; Thouez, 1973). People's residential address reflects, in a general sense, the basis of a social stratification of urban space and this social segregation is defined in relation to a number of different factors such as education and economic levels, ethnic origins, family size and so on. In France as in England, as well as in North-America, the main statistical tool has been socio-economic census data (Desplanques, 1976; Fox; 1977, Kitawaga and Hauser, 1973).

Socio-economic categories, however, cannot be viewed as identical to the concept of social class. Nonetheless, whatever measurement tool is used, the spatial analysis of differential mortality in its relationship to spatially segregated social stratification almost invariably reveals as inverse relationship between risk of death (in probability terms) and the social hierarchy. Furthermore social class in mortality rates do not appear to diminish over time (Surault, 1979).

As early as the end of the 18th century and especially into the nineteenth century health researches (doctors, demographers, statisticians...) noted mortality differences among the children of wealthy and poor districts of Paris. Housing and differential health conditions were linked to infant mortality thus beginning the quest for understanding of the relationship between health and mortality and urban social space (LecLerc 1979). Villerme notes, for example, in the Annales Generales de Medecine, that there was a strong correlation between the poor and high mortality rates in the 12 arrondissements of Paris during the 4-year period, 1817-1821. He took as indicator of health, deaths in the home, and as indicator of poverty, non-taxable rentals. Elsewhere, in England during the 1830's Chadwick's **Sanitary Report** also showed above-average mortality rates among the disinherited. Chadwicks further developed the miasmatic theory, fashionable at the time, whereby disease causing "miasmes" were considered to be generated by organic detritus and to be particularly responsible for cholera. However, in 1848 John Snow, using mapping techniques, brillantly demonstrated that cholera is transmitted by water and by the end of the century, both Pasteur and Koch had definitely laid to rest the so-called miasmatic theory. It should be mentioned, however, that even if this conceptual framework had a negative effect in the short term, it had the advantage of attracting attention to the problem of insalubrious conditions in homes and streets of economically depressed areas of the cities of that time.

The second half of the 19th century was characterized by withdrawal or an absence of statistical research on the spatial distribution of mortality. However, the first half of the 20th century was marked in France by an increasing interest in the comparative analysis of death rates according to the socially differentiated districts of certain cities (Felvay and Auvenque, 1953). In England and Wales, similar developments in mortality data analysis were taking place, with a special emphasis on profession related differential mortality. The same type of frame of reference was used by and large in work carried out in the Netherlands, Italy and the United States (Slidman et al., 1962; Class, 1964; Federici, 1969). One of the problems with using professional attachment as a risk factor related to death rates is that as age advances, there is often a certain modification of a person's professional profile.

The contemporary study of differential mortality in France is largely based on the observation over time of death rates of a national population sample selected in relation to the 1954 census (Desplanques, 1976). The main thrust of this work shows important geographic differences in infant mortality and in life expectancy at birth according to socio-professional categories. In the United States, the relating of death certificates and census data resulted in the compilation of a population of 340,000 deaths 1960. (Kitagawa and Hauser, 1973). Chi-squared analysis relating overall mortality to income-levels, educational levels and professions showed once again the inverse functional relationship between socio-economic status and death rates. In England, studies by Fox (1977) have shown the same thing, with the difference that the spread of mortality rates between professions is less marked than in the previous case (i.e., Kitagawa and Hauser, 1973). It would seem in this case that the mortality rates for middle and high income groups are more similar than in the U.S.A. Obviously this may well be partially explained by methodological differences between the two studies. One thing remains certain out of the comparison between the British (Fox) and American (Kitagawa and Hauser) studies: the poor have a much greater chance of dying early that those who are economically more comfortable. Finally, age-group and sexes being equal, significant inter-regional differences within countries have been noted. Wales, the north of England, the southern United States, Britanny, Alsace and Limousin in France are clear examples of regions of much higher than national average mortality rates (Surault, 1979).

In addition to the socio-economic factor (particularly when measured by professional activity) as an explanatory variable in predicting mortality rates, two other dimensions have often come into focus: environmental conditions (physical and social) and individual life-styles. In the first case, a direct relationship has been observed between increases in mortality rates for certain causes of death and city-size (Basavarajappa and Lindsay, 1976) particularly in the case of central planned-economy urban agglomerations (Desplanques, 1976). On the other hand, in the case of life-style variables it is known that mortality rates are higher for unmarried than for married people (Statistics Canada, 1980). It is also well established that consumption of medical service varies considerably with people's self-perception of their illnesses. Ecologically speaking, these types of phenomena can be grouped into life-style and behavioural classes. Furthermore, the whole question of perception in the geography of health and disease also fall into this category of predictor variables.

Insisting on the socio-cultural nature of mankind's inequality in the face of death and disease, although focussing on a real problem, does not resolve the question. The real heart of the matter is that health and disease are but one component of the larger social problem of the inequality of opportunity which besets the human race. Sauvy (1962) sums it up when he says **l'inegalite devant la mort, qui est inegalite devant la vie, apparait comme la resultante de toutes les inegalites.** In the same vein, research results obtained in western countries have shown conclusively, notwithstanding the difficulties of adequate observation and measurement, that the highest death rates are consistently found in the low level of the economic spectrum. Working conditions, life-style factors, environmental variables are also often found to be related, both to mortality and morbidity rates. In most of the preceding studies the role of geographic space was generally viewed indirectly. However, since social variables generally inscribe themselves geographically through spatial segregation, the above-mentioned explanatory variables always have some geographic dimension. This spatial dimension is illustrated by the following case study.

2. Case study: Mortality and Morbidity Rates on Montreal Island

As already noted, two methodological steps can be distinguished in geographically-based correlation studies linking population and environmental characteristics with either mortality or disease. The first stage is that of mapping the disease in question, and then the choosing of appropriate potentially causal ecological factors following this hypothesis - generating stage. And finally in this first analytical phase is the quantitative analysis through various statistical and mathematical techniques that is necessary in order to produce the various indices and subsequently the correlations between illness or mortality and the various postulated explanatory variables (Pyle, 1979). The second analytical step is that of defining **a priori** socio-economic and cultural areas sufficiently homogeneous that one can subsequently study the statistical differences in rates of morbidity or mortality in different zones (Loslier, 1976). The idea at this stage is to determine homogeneous regions where rates vary minimally within the zones and maximally between the zones. This chapter presents an example of the first type of analytical approach.

(a) The Data

The morbidity and mortality data for Montreal Island come from original comilations provided in the first case by the Departments of Community Health on Montreal Island (Mongeau et Lescarbeault, 1980) and in the second case by Statistics Canada.

The morbidity data are based on the frequencies of illness diagnoses in short-stay hospital centres between January 1st, 1975 and December 31st, 1977. They are drawn from standard hospital admission-release forms (AH101). This information was spatially aggregated at the level of C.L.S.C. (Centres locauxde sante communautaire) zones. The mortality data are based on the total set of deaths in the Montreal area during 1971. Using the residential addresses once again, deaths were grouped by census tract and then by C.L.S.C. zone. The C.L.S.C. is a primary health care unit serving a specific area of the city. There were 41 C.L.S.C. areas on Montreal Island in 1976. For the purpose of this paper it was decided in view of their relative importance to retain all diagnostics concerning cardiavasculaire illness (CIMA 460-519). Hospital morbidity and mortality data showed that in 1976 these types of diagnostic accounted for 9% of all hospitalizations and for 47% of all deaths.

(b) The Method

The methodological approach in this illustrative case study is divided into three stages. First, the mapping of comparative indices of hospital morbidity and of mortality. The mortality indices were calculated by means of the indirect method of standardization (Jenicek, 1976) taking the population of the Montreal C.M.A. as the reference population. This technique permits us to remove from the data the effects of age and sex, or in other words, holds these latter confounding variables constant. Furthermore, standard normal - scores were calculated and mapped enabling us to visualize the position of each area in relation to the mean values and the dispersion of the mortality and morbidity data.

The second step involved the principal components analysis of the socio-economic, population and geographic characteristics of the 41-zone geographic analysis grid. The socio-economic variables included an index relating educational level and income (Carlos index), percentage French-speaking and average numbers of occupants per room. For the population characteristics, variables included were an aging index, median age, population growth rates and, finally, percentage of those 14 years of age or less. The term geographic variables refers to density and mobility measurements, as well as a constructed index of geographic accessibility to the hospital network. The goal of the multivariate analysis of these data was to eliminate redundancy and multicollinearity among the independent potentially explanation variables.

Finally, the third and classic step was to conduct multiple regression analysis in which the dependent variable is either morbidity or mortality by cause (diagnostic grouping) and the independant variables are the most important variable (in terms of factor loadings) of each of the six principle components obtained from the previously mentioned analysis. (See table 13.1 for the list of variables with their definitions.)

(c) Results

Maps 13.1 to 13.4 enable us to visualize normalized distributions of both morbidity and mortality in terms of cardiocascular disease on Montreal Island. The distribution of high rates of morbidity for men (Map 13.1) takes the form of a crescent running from the Lachine C.L.S.C. (See reference map - Map 13.5) on the west to the Ahuntsic C.L.S.C. in the north central section of the island, passing through downtown districts such as St-Jacques. The geography of cardiovascular mortality for men (Map 13.3) is much less clearly structured in a spatial sense. The highest rates of this type of mortality are found in the poorer districts of the centre south, St. Henri and Pointe St.Charles, for example (See Map 13.5 for reference purposes) as well as in strongly francophone areas such as Mercier Est and Ville d'Anjou. The geographic distribution of high levels of morbidity for cardiovascular disease in women (Map 13.2) follows closely the map for men (Map 13.1) but the female mortality map (Map 13.4) differs to a certain extent from the male mortality map. Both men and women suffer from high mortality rates in the poorer districts of the Centre-south of the island, but high female mortality seems to be concentrated in the francophone and ethnic districts in the Northeast of the Island - St-Leonard, St. Michel du Nord (See Map 13.5).

TABLE 13.1
List of Independent Variables

Socio-economic variables

1. The Carlos socio-economic index resulting from the combination of educational and income status through a factor analytic procedure.
2. Per cent French-speaking in relation to all tongues used in each zone.
3. Number of people per room, 1971.

Demographic variables

4. Aging index: the ratio between those over 65 and those under 15 years of age in 1976.
5. Median age: by C.L.S.C. zone in 1976.
6. Growth rate: expreseed as a per cent, corresponds to the ratio between population change (pop. 1976 - pop. 1966) divided by the 1966 population for each C.L.S.C. zone.
7. 0-14 years: all members of this age-group in 1976.

Geographic variables

8. Density: population density per km^2 in 1976.
9. Mobility - non-movers: those in 1976 having lived in the same residence 5 years or more.
10. Mobility - within Quebec: those in 1976 having moved within the province in the past 5 years.
11. Mobility - from outside Quebec: those in 1976 having moved **into** Quebec over the last 5 years.
12. Accessibility to the hospital network: this index was developed by summing the weighted distance ratios for each zone to the 3 closest hospitals. These hospitals were weighted in relation to the size of their respective clienteles.

Table 13.2 presents the results of the principal components analysis. The first component represents age structure with loading of -0.91 for the 14 and under age-group and +0.86 for the over 65 years age-group. The second component can be summarized under the label of socio-economic characteristics. The highest loadings on this component are -0.86 for number of people per room and +0.79 for the Carlos index, a combination

FIG. 13.1

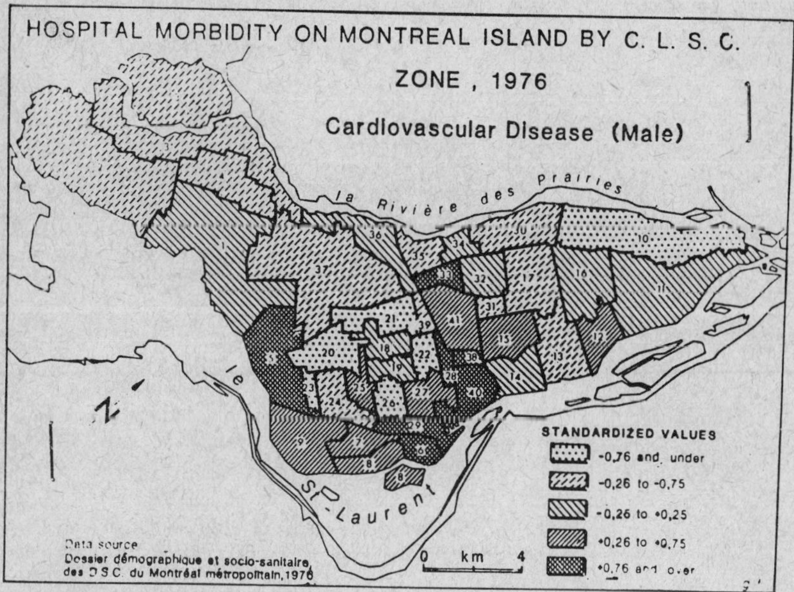

FIG. 13.2

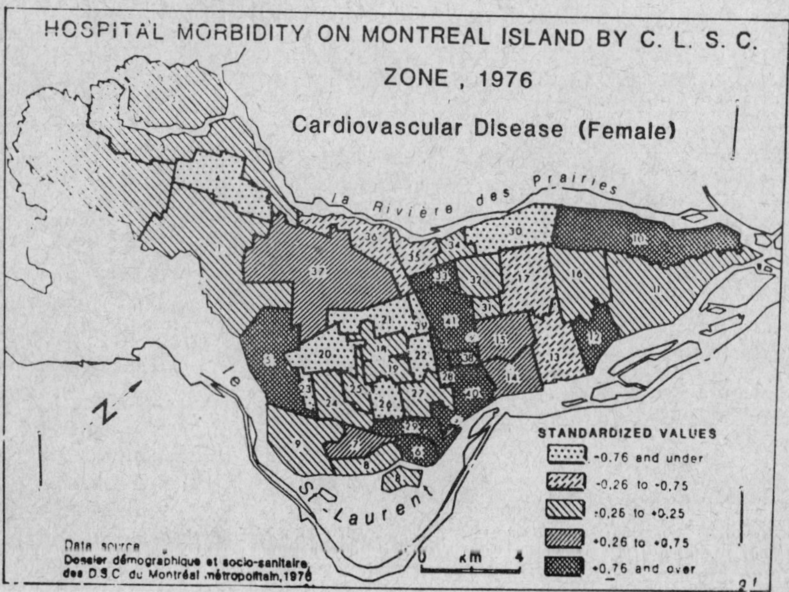

FIG. 13.3

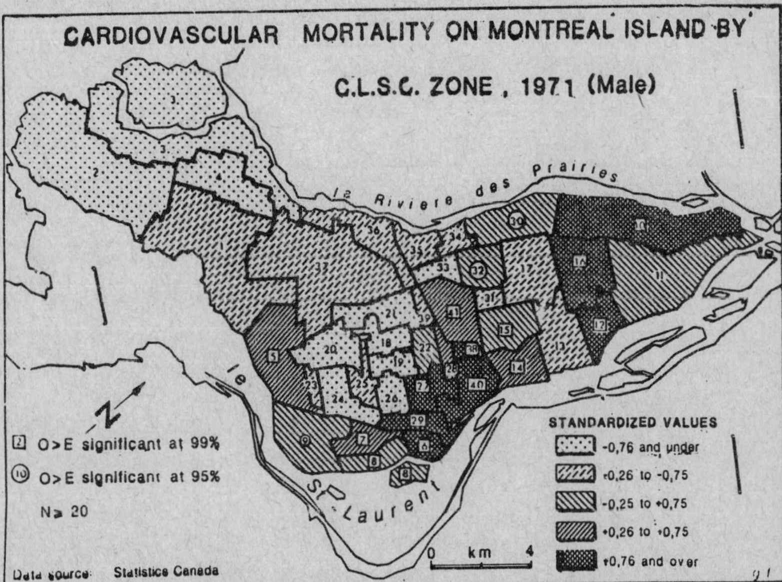

FIG. 13.4

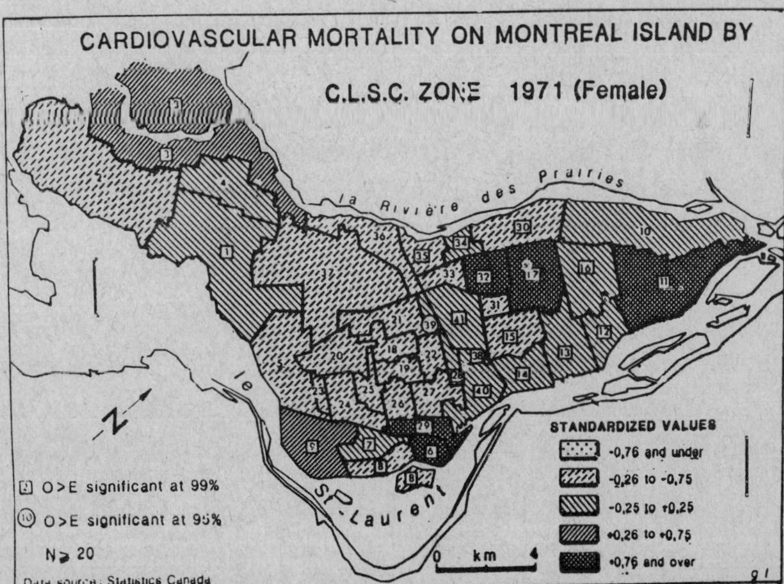

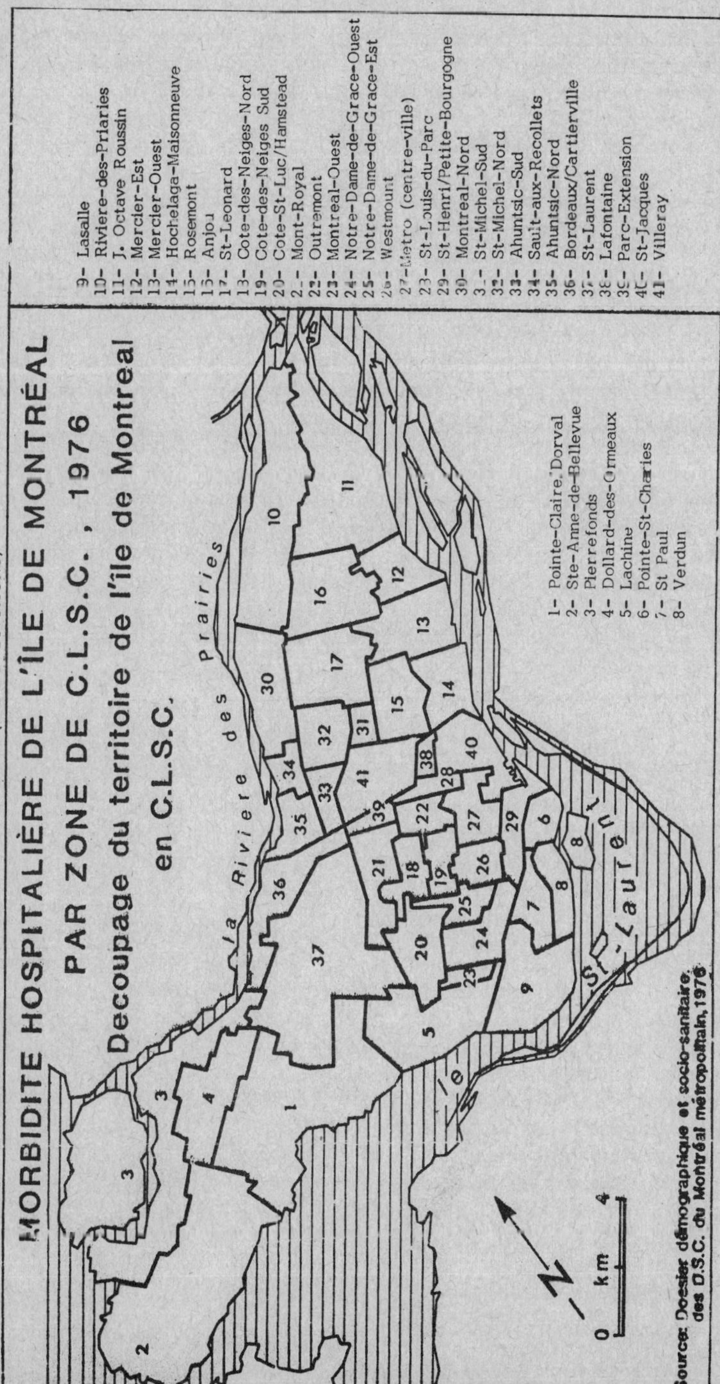

FIG. 13.5 - REFERENCE FIG OF MONTREAL ISLAND (C.L.S.C. DIVISIONS).

of educational and income levels. The third component represents mobility characteristics, the highest loadings being +0.84 for population growth rates, -0.81 for people having stayed in the same dwellings between 1971 and 1976 and +0.75 for people having immigrated from outside of Quebec or Canada. The fourth component is related to ethnicity, the highest loadings being -0.71 for percentage of French-speaking people and +0.90 for those moving into the province. The fifth and sixth components present respectively isolated loadings of 0.91 for population density and 0.92 for accessibility to hospital services. These last two components sum up in a non-redundant way the geographic variables.

For the regression analysis the variable showing the highest meaningful factor loading for each component was used as representative of that component. As can be seen from Table 13.3, hospital morbidity is clearly but inversely linked to socio-economic status; both for men and women. This relationship goes in the same direction when mortality data is used in this type of regression, but the relationship is not statistically significant in the case of men (Table 13.4). Conversely, residential mobility is statistically significant in the prediction of male mortality, whereas residential stability predicts significantly for woman only, as is álso the case for the age index (Table 13.4).

(d) Comments

The results for Montreal Island confirm the close relationship between socio-economic status and cardiovascular disease. Loslier (1976) showed by another methodological approach that the comparative mortality index (C.M.I.) for ischemique heart disease for both sexes for the 35-64 age group was 1.47 in the centre-east of the island where most of the lower income areas are found, while more economically secure areas had an overall C.M.I. of 0.60. We have seen here, however, that socio-economic factors include social and physical environmental factors (including working conditions) as well as behavioural questions such as life-style, diet, etc... Almost all these factors work together synergetically and it is for this reason that it is so difficult to single out specific causal factors. Table 13.5 suggests possible links between predictor variables and cardiovascular disease. For most of the relationships the actual mechanisms linking the variables to the genesis and development of heart disease remain to be discovered. For the time being we can use socio-economic status and its components as what might be called "persuasive" factors; that is to say, they are factors which, while not necessarily the direct cause of disease, control to a large degree the extent to which disease may be resisted.

TABLE 13.2
Principle Component Structure

Components	Eigen-Values	Pct Expl. Var.	Pct. Cum
I	4.04	33.7%	33.7%
II	3.32	27.7%	61.4%
III	1.61	13.5%	74.9%
IV	1.09	9.1%	84.0%

Variables / Components	Socio-Economic Index	Persons per room	% French language	Age Index	Median Age	Growth Rate	% 0-14 Years	Density	Mobility: non-movers	Mobility from-within Quebec	Mobility: from outside Can. & Que.	Accessibility to hospital
I	0,14	-0,32	0,06	<u>0,86</u>	0,72	-0,27	<u>-0,91</u>	0,21	-0,16	-0,31	0,12	0,3
II	<u>0,79</u>	<u>-0,86</u>	-0,60	0,07	0,48	0,12	-0,13	-0,23	0,25	0,31	0,20	0,0
III	0,16	-0,02	-0,01	0,01	-0,22	<u>0,84</u>	0,24	-0,18	<u>-0,81</u>	<u>0,75</u>	0,16	-0,08
IV	0,25	-0,21	<u>-0,71</u>	0,23	0,13	0,06	0,12	0,10	-0,23	-0,10	<u>0,90</u>	0,11
V	-0,43	0,11	0,02	0,02	0,21	-0,12	-0,13	<u>0,91</u>	-0,02	-0,27	0,09	0,05
VI	-0,06	-0,06	-0,12	0,34	0,13	-0,13	-0,09	0,04	-0,21	-0,17	0,07	0,92

TABLE 13.3

Montreal Morbidity Regression Analysis - Cardiovascular Disease

	Socio-economic Index	Accessibility to hospitals	Non-movers	Mobility from outside Quebec and Canada	Age Index	Population density	Total (R^2)
MEN							
Level of significance	0,005*	0,671	0,655	0,692	0,275	0,552	0,031
Cumulative R^2	28	29	29	29	31	32	32
	-0,676	-0,079	0,072	0,078	0,225	-0,124	
WOMEN							
Level of significance	0,004*	0,605	0,828	0,635	0,792	0,885	0,007
Cumulative R^2	37	38	38	38	38	38	38
	-0,656	0,091	0,033	0,089	-0,051	-0,028	

TABLE 13.4

Montreal Mortality Regression Analysis - Cardiovascular Disease

	Socio-economic Index	Accessibility to hospitals	Non-movers	Mobility from outside Quebec and Canada	Age Index	Population density	Total (R^2)
MEN							
Level of significance	0,867	0,816	0,329	0,039[x]	0,748	0,351	0,178
Cumulative R^2	11	11	11	18	19	22	22
	-0,589	-0,466	-0,170	-0,449	0,210	-0,752	
WOMEN							
Level of significance	0,056	0,680	0,022[x]	0,787	0,001[xx]	0,164	0,001[xx]
Cumulative R^2	21	21	36	36	50	53	53
	-0,365	0,638	-0,319	0,442	0,581	-0,244	

One other important observation has to do with the geographic location of high morbidity and mortality rates. Areas that have high rates (therefore, lower health levels) tend to be concentrated in high density, industrial areas. This in turn places the populations of such zones at risk from high levels of air pollution (Lavergine et al., 1975; Holland and Reed, 1965). Finally, the fact that areas of high mortality and morbidity rates are concentrated spatially represents a practical advantage in the implementation of appropriate preventive health measures and programmes.

3. Limits to the Geographic Approach in the Study of Disease and Mortality.

The limitations to this type of study are well known but sometimes overlooked. One potential problem is that of "false positives" and "false negatives" (Blalock, 1961) and their importance in the context of geographic pathology. The "false positive" result can occur when one transposes a group observation to the level of the individual. This is an aspect of what is sometimes called the "ecological fallacy". To illustrate, imagine a correlation study based on geographic units (e.g. counties, states or provinces) where a positive correlation has been found between exposure to dust and lung cancer. If in the geographic units where there is high dust exposure, there is also a high level of tobacco consumption, the apparent relationship between dust and lung cancer might disappear when one studies the risk factors at the level of individual cases through a valid epidemiological approach. What happens, in effect, is that operating at the group level (geographic units) the potential confounding factors are greatly increased.

The "false negative" result, sometimes called the "atomistic fallacy" can occur in two different ways. Firstly, it could involve a confounding negative by which, say for example, the causal association between occupational exposure to lead and lung cancer could be masked or blanked out in terms of a geographic correlation analysis. This could happen, for example, if the geographical units where foundries are located are also the ones where tobacco consumption is lower than in zones where there are no foundries. Secondly, the "false negative" result can occur as a result of a diluting mechanism: for example, a known association at the individual case level between occupational exposure to epirite and lung cancer could be lost at the level of geographic correlation because, even in the absence of true confounding variables, the number of cases of lung cancer due to work exposure is very low in relation to the high number of cases in the population attributable causally

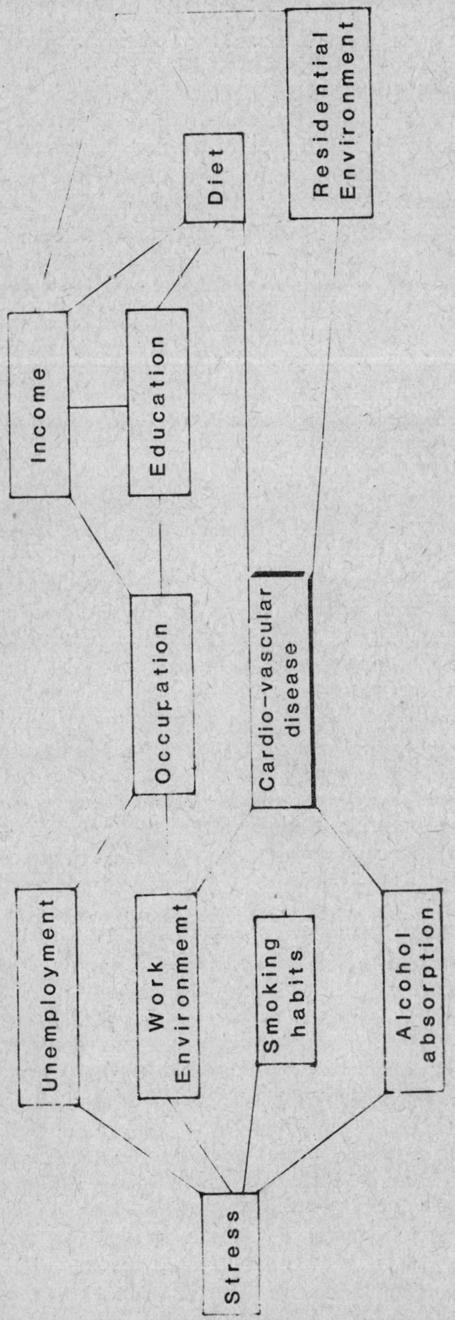

TABLE :13.5
Associations between cardio-vascular disease , socio-economic variables , life-style and environment

to tobacco consumption.

Cleek (1979) has raised some of those problems in relation to the effect of using different geographic scales on the interpretation of research results. Taking up the simple but precise approach of Jansson (1978) he notes that a high coefficient of variation (standard deviation expressed as a percentage of the mean) for a given cancer site, at a given regional scale suggests "environmental dependance on some factors which vary at the same scale as that on which the cancer rate is aggregated. A low coefficient of variation need not suggest a lack of environmental dependence, but instead little or no dependence on an environmental factor which varies at this scale". Furthermore, since the coefficient of variation is calculated on the basis of the mean and the standard deviation, it has little meaning if the data are not distributed normally. This is the case when there are few cases involved. However, in the case of important disease types,

> "changing measures of explained variation, obtained when data are aggregated at different scales, can themselves lead to hypotheses concerning the spatial scale at which environmental or socio-economic clues are operating. A final problem associated with the use of aggregated level data is attributable to spatial autocorrelation. Essentially, autocorrelation in the independent variables can reduce the power of inferential parameters. The effect of autocorrelation in the dependent variable, the disease rate, is less clear. Shafer (1978) finds that primacy bone cancer rates show different levels of spatial autocorrelation when mapped for countries, state economic areas and random aggregations of adjacent countries in Pennsylvania. He indicates also that identifiable peaks and lows in the disease rate surface are inconsistent when different country aggregations are used. These results have serious implications for geographical epidemiology and indicate clearly that much work needs to be done in this area, as in all areas of spatial analysis discussed above."
>
> (Cleeks, 1979, p. 244)

There is certainly the advantage with individual level data that the researcher is closest to the process of interest and the use of clustering techniques is a powerful tool to detect disease patterns. However, the advantages of aggregated data would seem to outweigh the use of individual level data, when they are available at different aggregation levels. Similarly, the extensive manipulation necessary to unravel the more complex disease relationships is facilitated by the use of aggregated data. It is not that one type of study is superior to the other but that they complement each other.

Even with these various methodological problems we are far from having exhausted all the valid possibilities of analysis in the geography of health and disease. If this rich and varied subdiscipline of geography which owes so much to connected disciplines such as epidemiology and medicine (Picheral, 1982) has not contributed more to the understanding of disease etiologies, it is not so much our lack of analytical techniques as it is our lack of capacity to conceive of valid hypotheses to explain particular types of spatial distributions.

CONCLUSIONS

These results based on Montreal data confirm previous research indicating that high rates of cardiovascular mortality and disease are closely associated with socio-economically depressed zones within urban space. This fairly large diagnostic grouping, including ischemic heart disease and other cardiovascular diseases all of which have differing etiologies, seems to be significantly related to high levels of hypertension, cholesterol and triglic rides (Brutin and Wolfe, 1970). These factors tendent to increase in the presence of prolonged stressful situations (Selve, 1971). In some way the physiological environment has to be related to a person's physical environment. Thus we conclude that some environments are "supportive" while others are detrimental to the physical well-being of an individual. It would appear that this is particularly true in the case of cardio-vascular disease.

Studying the psychological (cognitive) dimension of the social environment is perhaps a helpful approach. One of the most complete studies in this field has shown that in spite of the presence of other risk factors such as smoking, obesity and so on, the Italian-American community of Rosito, Pennsylvania has relatively low rates of heart disease and that this is due largely to their unstressful social environment (Moos and Isil 1974). This would seem to be one of the most promising avenues to explore in future studies in our efforts to understand the meaning of the demonstrated relationship between cardio-vascular disease and socio-economic variables.

REFERENCES

Basavarajappa, K.J. and Lindsay, J., **Variations de la mortalite au Canada,** Statistiques Canada, Catalogue 84 533, (1976).

Berry, B.J.L. and Kasarda, J.D., **Contemporary Urban Ecology,** MacMillan, (1977), 496 p.

Blalock, M., **Causal Inferences in Non Experimental Research,** University of North Carolina, Chapel Hill, (1961), Chap. 4.

Burgess, E.W., The Determination of Gradients in the Growth of the City Class, D.V. **Pubs. Am. Sociol. Soc. 21,** (1927), 178-184.

Bruhn, J.G., and Wolfe, J., Studies reporting "low rates" of ischemic heart disease: a critical review. **American Journal of Public Health,** 60, 192, 1477-1496.

Class, D.V., Some Indicators of Differences Between Urban and Rural Mortality in England and Wales and Scotland, **Population Studies,** march 1964.

Cleek, R.K., Cancers and the Environment. The Effect of Scale. **Soc. Sc. and Med.** Vol. 13D, (1979), 241-247.

Chadwick, E., **Report on the Sanitary Condition of the Labouring Population of Great Britain,** Flinn, M.W. (edt.) Edinburgh University Press, (1965) (1842).

Desplanques, G., **La mortalite des adultes suivant le milieu social, 1955-1971.** Collection de l'INSEE, serie M, 44, avril (1976).

Federici, N., **The Impact of socio economic factors on mortality.** An attempt at Analysis of Some Italian Data in Congres International de la Population, Londres, (1969), 2 vols.

Felvay, M. et Aubinque, M., La mortalite par categorie socio-professionnelle, **Etudes Statistiques** 3, juillet-septembre 1957.

Foggin, P. and Polese, M., **La geographie sociale de Montreal en 1971.** Institut National de la Recherche Scientifique - Urbanisation, (1976).

Fox, J., Occupational Mortality 1970-1972., **Population Trends,** Automna, (1977), 8-15.

Greer Wooten, B. The Urban Module in L. Beauregard (ed), **Montreal: Field Guide,** Presses Universitaires de Montreal, 1972.

Harris, C. and Ullman, E.L., The Nature of Cities. **Annals of the American Academy of Political and Social Sciences,** 242, (1945), 7-17.

Holland, W.W. et Reed, D.D., The Urban Factor in Chronic Bronchitis. **The Lancet,** 27, February 1965, 445-448.

Howe, M.G. and Loraine, J.A. **Environment-medicine,** Heineman, London, (1976), 263p.

Hunter, J.M., "Strategies for the Control of River Blindness", Meade, M.S. (ed.) **Conceptual and Methodological Issues in Medical Geography,** Studies in Geography 15, Department of Geography, University of North Carolina at Chapel Hill, 38-76.

Janson, B., Seibert, G., Speer, J., "Gastrointestinal Cancer". **Cancer,** N.Y., 36, (1975), 2376.

Jenicek, M., **Introduction d l'epidemiolgie,** Edisum., St. Hyacinthe, Quebec, (1976).

Kitagawa, E., et Hauser, P., **Differential Mortality in the United States: a Study in Socio—economic Epidemiology.** Cambridge, Mass., Harvard University Press., (1973).

Lavergne, B., Frappier-Davignon, L. St. Pierre, J., Etude de l'effet de la pollution atmospherique sur la mortalite par cancer du poumon et par cancer de l'intestin., Ile de Montreal, (1963-1966), **Union Medicale du Canada,** 1397-1406.

Leclerc, G., **L'observation de l'homme.** Le Seuil, Paris, (1972), Chap. 9.

Loslier, L., **La mortalite dans les aires sociales de la region metropolitaine de Montreal,** Direction Generale de la Planification, Service des Etudes Epidemiologiques, Ministere des Affaires Sociales, (1976).

May, J.M., Medical Geography; its Methods and Objectives., **Geog., Reivew,** 56, (1966), 398-416.

Meade, M., Medical Geography as Human Ecology: The Dimension of Population Movement., **Geog. Review,** 67, 4, 379-395.

Mongeau, J., et Locoarboault, G., **Dossier** demographique et socio-sanitaire des Departements de Sante Communautaire du Montreal metropolitain. Etude commandee par les D.S.C. de la region 06-A l'I.N.R.S.-Urbanisation, 2 vols., novembre 1980.

Moos, R. and Insil, P. **Issues in social ecology, human milieus,** Polo Alto: National Press Books, (1974).

Park, R.E., Burgess, E.W., and McKenzie (eds) **The City,** Chicago (1925).

Picheral, H., "Geographie medicale / Geographie des maladies / geographie de la sante", **Espace Geographique,** Vol. XI, No. 3, (1982), 161-175.

Pyle, G., **Applied Medical Geography,** J. Wiley, (1979), Chapitre 3.

Pyle, G., **Heart Disease, Cancer and Stroke in Chicago: A Geographical Analysis with Facilities, plans for 1980,** Research Paper No. 13, Department of Geography, University of Chicago (1971).

Sauvy, A., **Les Limites de la vie humaine,** Hachette, Paris (1962).

Seidman, M., Garfinkel, L., and Graig, L., Deaths Rates in New York City by Socio-economic class and religious group and by country of Birth, 1949-51, **Jewish Journal of Sociology,** 4, december 1962.

Selye, H., The evolution of the stress concept in Levi, L. ed. **Society, stress and disease,** Oxford University Press (1971).

Shafer, S.Q., **Mapping Mortality: The Search Peaks of Bone Cancer Death Rates in Pensylvania counties.** Unpublished M.A. Thesis, Department of Geography, Columbia University Press, (1978).

Shevky, E. and Bell, W., **Social Area Analysis: Theory, Illustrative Application and com tional procedures,** Stanford Univer-

sity Press (1955).
Sorre, M., **Les fondements biologiques de la geographie humaine: essai d'une ecologie de l'homme.**, A. Colin. (1951), 447 p.
Statistics Canada., **Vital Statistics**, vol. III, Deaths Catalogue 84-906, Annual, Ottawa, 1980.
Surault, P., **L'inegalite devant la mort**, Economica, Paris (1979).
Thouez, J.P., La structure spatiale des caracteristiques socio-economiques de Montreal, 1961, Partie I, Rotation orthogonale versus rotation oblique., **Recherches Sociographiques**, 14, 1, (1973), 81-115.
Villerme, L.R., **Etat physique et moral des ouvriers.**, TYL, Y. (coll) Union Generale d'Editions, Paris, (1971) (1840).

FOURTEEN

Derek R. Hall

Emerging Residential Structures of Socialist Cities

SOCIALIST SETTLEMENT PLANNING STRATEGIES

While settlement planning in socialist societies has pursued a number of common goals, the cultures and environments within which such policies have been attempted have been remarkably diverse. The combination of centralised economic planning and public control of the means of production has permitted large scale transformation to take place in such societies, although individual national differences in pre-existing settlement forms, demographic structures and levels of economic development have varied the impact of those transformations from one country to the next.

In most socialist societies, settlement planning is set within national economic development plans and industrial and agricultural regional strategies, such that two general aims have been recognised: the attainment of a planned and balanced settlement system; and the establishment of a rational distribution and use of productive forces (Musil, 1981).

These general aims have been supplemented by more specific objectives:

(i) The almost paradoxical attainment of a balance between social equity and economic efficiency in spatial development;

(ii) reduction of social inequalities between regions, by aiding backward areas, and improving the distribution of, and access to, public facilities and services;

(iii) the integration of urban and rural settlements through:

(a) an emphasis on strong regional centres and rural settlements concentration;

(b) the provision of non-agricultural employment in rural areas;

(c) the introduction of urban standards and levels in housing, services and other rural facilities;

(iv) a retardation of the growth of large agglomerations to facilitate a more rational spatial distribution of production, transport and housing;

(v) an improvement in environmental quality.

Such principles and objectives have, inevitably, been subsequently modified. Initial emphases upon equality of population density and settlement distribution, with an equal allocation of public facilities and services, have largely given way to the requirements of economic efficiency and rapid improvements in living conditions, reflecting a shift from normative versions of central place theory to growth pole and agglomeration strategies. Musil (1981) has recognised this shift in a number of trends:

(i) increased emphasis on the economic role of settlement structures within the context of achieving improved national economic efficiency;

(ii) the rational concentration (rather than dispersal) of economic and social activities, implicitly on grounds of 'efficiency' rather than equity;

(iii) the planned formation and regulations of large agglomerations, most notably in Czechoslovakia, the German Democratic Republic, Poland and the Soviet Union;

(iv) a greater flexibility in the employment of indigenous settlement forms, as, for example, Yugoslavia, Romania and China;

(v) national variations in attitudes towards 'optimum' degrees of concentration, settlement sizes, levels of urbanisation, etc.

Within the context of rapid change at both the national and international levels, each socialist society now needs to balance out the often conflicting economic, social and ecological effects of urbanisation, contemporary levels of which can be seen in table 14.1.

Generally, several phases of urban development in the period since World War II—the period within which most socialist societies took their new ideological guise—may be identified:

(i) in the immediate post-war periods,[1] highest priority was given to rebuilding war damaged cities and industry, adapting both to a new socialist framework and often revised international boundaries;

(ii) mass housing programmes followed, to meet the needs of the new urban proletariat whose numbers were being raised by rural to urban migration and by (usually encouraged) increases in the birth rate;

(iii) a reduction of slum and squatter settlements;

(iv) a more recent phase of city centre reconstruction in a move back to large scale central retailing or the incorporation, as in Romania and North Korea, of indigenous cultural elements rather than austerely symbolic 'socialist' architectural styles.

Socialist urban spatial structure

Socialist policies have usually had to be superimposed upon structures and patterns which had evolved under very different political conditions. Only in socialist new towns and rebuilt water-destroyed cities have such policies had a full rein, and even in the latter case constraints have been imposed by the need to conform to surviving infrastructural elements, such as water supply and sewage systems, and by the rebuilding of important cultural and national symbols such as Dresden and Old Warsaw. The first socialist urban planning 'blueprint' was the Moscow Plan of 1935, which brought together a number of trends that had been emerging in the Soviet Union since the 1917 revolution. Although the plan's main principles were to be often dogmatically followed throughout the subsequently expanding socialist world, latterly, modifications to, and even reversals of, some of those guidelines, have come as no surprise. In particular, paradoxes and contradictions surrounding the questions of optimum city size, interpretation of spatial equality, and control of the housing market, have helped to mould, define, and re-define the spatial nature of internal urban structure in socialist societies.

TABLE 14.1

Recent urban growth in the socialist societies

Country	% urban pop.			Urban pop.an av. growth rate		% urban pop. in largest city		% urban pop. in 1/2 m+ pop. cities	
	1939	1960	1980	1960-70	1970-80	1960	1980	1960	1980
Albania	15	31	37*	3.7	3.4	27	25	-	-
Bulgaria	25	39	64	3.8	2.6	23	18	23	18
Cuba		55	65**	2.9	2.4	38	32	38	32
Czechoslovakia	33	47	63	2.1	2.0	17	12	17	12
D.P.R. Korea		40	60	5.0	4.3	15	12	15	19
German Dem. Rep.	72	72	77	0.1	0.3	9	9	14	17
Hungary	35	40	54	1.7	2.1	45	37	45	37
Mongolia		36	51	5.2	4.1	53	52	-	-
Poland	27	48	57	1.8	1.7	17	15	41	47
P.R. China		++	13++	++	3.1	6	6	42	45
Romania	23	34	48	2.8	2.5	22	17	22	17
U.S.S.R.		49	65	2.8	2.2	6	4	21	33
Vietnam		15	19#	5.3	3.3#	32	21	32	50
Yugoslavia	20	28	42	3.2	2.9	11	10	11	23

Sources: World Bank (1980) 148-9, (1981) 172-3.

Notes :In several cases the individual states' own statistics differ from the above, e.g.,

* 33.5 per cent (Papagjorgi, 1982)
** 69 per cent (1981 census, Granma, Nov. 8, 1981)

Several notable revisions in the World Bank's 1981 report contrasted with the figures of the previous year, e.g., (1980 report):

++ 1960 19 per cent 1980 25 per cent 1960-70 3.6 per cent
\# 1980 23 per cent 1970-80 5.1 per cent

Optimum city size

Representing both the spatial and social concentration of wealth, services and opportunities, large urban centres were viewed as symbolising capitalism in all its worse excesses. Under socialism, by contrast, opportunities and development would be equalised to constrain the growth or even reduce the size of existing urban agglomerations and preclude the emergence of further large centres.

In Czarist Russia, the growth of large cities such as Moscow and St. Petersburg could be retrospectively interpreted as manifestations of capitalist development. These centres' rapid growth during the 1920's and 1930's (the latter city having been renamed Petrograd and then Leningrad), long after the Bolshevik Revolution, constitued a number of urban planning dilemmas. Under the 1935 plan's guidelines, Moscow itself was to grow from a population of 3.5 million to five million (it is now eight million). Paradoxically, the ideal socialist city size was considered to be 50,000 - 60,000 population. This level would be sufficiently large to allow an effective functioning of services and economic provision of facilities whilst still small enough to permit the development of a communal socialist ethos. Arguing that above a certain critical threshold level, transport, utility, public health and other infrastructural costs rise more rapidly per inhabitant, a number of Soviet economists and architects[2] used housing, transport and other services' costs to evolve optimal urban population size ranges as wide as 50,000 to 250,000. Criticisms point to the fact that major economies of scale appear not to be taken into account in such formulations, while counter arguments point to the inefficiency and cost resulting from the congestion that large concentrations are likely to promote.

In the Soviet Union, standards and norms for urban planning and construction were issued in 1958 to impose further constraints on urban centres larger than 100,000 population, and to direct new industrial construction to cities of below that size. Subsequently, at least Soviet urban planners recognised the problem of optimal urban development to be less related to size than to an appropriate internal urban structuring. The 1968 increase in number of Moscow's local soviets, from 16 to 27, for example, reflected a recognition for the need for smaller, more effective, and balanced city planning units.

As one way of controlling urban growth, an internal passport system was re-introduced into the Soviet Union in 1932,[3] restricting the right of urban residence to those with proof of regular employment there. Similar devices were later employed in other socialist societies to constrain rural to urban migration. But at least in the USSR and Eastern Europe ever greater proportions of these societies' citizens are living in the largest cities, such that, superficially at least, the notion of an 'optimum size' would appear to be redundant, although some writers still give it lip service. As noted earlier, there has been a fairly drastic series of revisions to urbanisation policy in such socialist societies as the German Democratic Republic since the 1960's, whereby planners now argue that infrastructural costs are relatively higher for smaller urban places, such that new apartment construction has tended to be concentrated in the larger urban-industrial centres[4] to encourage population movement from the countryside to those cities where labour is most needed.

The restriction of large city growth, which appeared in the Soviet Union in the 1920's and 1930's, only became apparent in China a decade after the 1949 revolution. The first five year plan period (1953-7), was characterised by grandiose ideals of demolition and reconstruction, particularly in large cities. After 1960, however, the Chinese press blamed this earlier 'giganticism' on Soviet influence. Further development of large urban centres was curtailed, while the number of towns and cities of fewer than 100,000 inhabitants increased considerably.

Public Ownership of all urban land

Under capitalist-based urban development, land uses tended to be determined by land values which are related to site accessibility. While accessibility means different things for different functions, the city centre, theoretically the most accessible location from all other points within the urban area, reveals the highest land values, and is thus dominated by the highest paying commercial activities, which, in their turn, usually represent the greatest vertical development. The concentric, sectoral and nuclear elements of urban structure in the western world thus express the varied interrelationships between such dynamic factors as the land value surface, rent paying ability, historical development and functional segregation.

In the socialist city, by contrast, the public ownership of land removes any land value surface and land speculation. While urban authorities act as the resource allocator and general planner, as well as land owner, it has been argued that the absence of explicit land values results in an underuse of urban land, as also does land allocation based upon the expectation of future plans which may never bear fruition. However, after economic reforms of the 1960's in the USSR and Eastern Europe, recognition of such relative inefficiencies stimulated a search for suitable land value surrogates. One notable study (Kabakova, 1973) suggested the development of such a surrogate from the employment of fifteen variables. These fell into two groups—'socio-economic' and 'engineering-construction'—employing a quantification of such elements as accessibility, level of consumer and cultural services and quality of housing. Resulting internal regionalisations of Moscow, Leningrad and Yerevan-saw city centres receiving 'values' of up to four times those of peripheral locations, presenting patterns of concentric spatial zones decreasing in 'value' outwards. While large elements of subjective appraisal are evident in such an approach, its purpose, in attempting to provide an (ideologically acceptable) vehicle for greater discrimination in land use allocation processes, is significant in trying to provide a surrogate for market forces and to build the consequences of that into decision making processes (Bater, 1980).

Stringent land use zoning and rationalised internal movement

Commensurate with public control, stringent land use zoning to segregate disharmonious functions is officially pursued. While, for example, relative ease of access should be rendered from residential to industrial employment areas to reduce commuting time and costs, and congestion, both areas need to be sufficiently segregated to prevent industrial pollution adversely affecting residential areas.

Forty minutes was designated in the Moscow Plan as the maximum commuting time acceptable for a large city. However, by minimising journeys to work, close functional relationships between particular residential and employment areas could result in explicit residential segregation, particularly between 'white collar' employees working in city centres or particular 'wedges' of the city, and manual workers in factories, on construction sites or in other 'wedges'.

Generally, industrial and through traffic should avoid residential areas, routeways being designated in relation to levels of use, noise and atmospheric pollution, in order to regularise traffic flow and improve residential environments.

Extensive green open space should complement industrial zones. Theoretically, this allows more recreational opportunities than in capitalist cities, and provides buffers, green wedges and belts to cordon industrial areas and help restrict outward physical urban growth.

The nature and role of city centres

An early ideal set out for the internal structuring of socialist cities was the elimination of any distinguishable centre: concentrations of consumer and commercial activity symbolised capitalist development and encouraged congestion. By the late 1920s however, the notion had gained favour that a city centre should be present, but only to act as the nucleus of urban social and political life, and not for economic activity. 'Socialist' architecture—initially Stalinist monumentalism—would buttress such ideals. Retailing and other commercial activities would be decentralised into residential neighbourhoods, with industry and warehousing concentrated in industrial zones.

Subsequent planning of socialist city centres—often entailing the remoulding of existing structural relationships—has therefore entailed making provision for a focus for socialist mass political and cultural activity: spatially and architecturally expressing the dominance of the party through institutional buildings, large squares for parades and processions, heroic monuments and propaganda hoardings. While many of these characteristics were inherited from pre-socialist regimes and can be found in city centres all over the world, the dominant association of the ruling party should distinguish city centres in socialist societies.

The low level of retailing in socialist city centres reflects not only positive decentralisation, but is also the result of the combination of public control of retailing activities and the lack of formal competition between outlets, lower levels of purchasing power and consumer goods production, and an emphasis upon national savings schemes. Consequently, there are fewer shops than would be found in a comparable western town or city. Under these circumstances, the term central cultural district (CCD), rather than central business district (CBD), has been applied to the theoretically non-commercial socialist urban centres.

The earlier monumentalist architectural style has, however, tended to give way to more indigenous elements, and forms familiar in the west, such as prefabrication. Particularly in the capital cities, the emergence of a prestigious 'modern' city centre is coming about, such as in Warsaw's Centrum shopping complex, or East Berlin's Alexander Platz.

Within this context, the emerging residential structures of contemporary socialist cities may be seen as the result of two interacting dimensions: the planning of residential areas and the changing nature of housing tenure.

Planned development of residential areas

The residential neighbourhood ('micro-district') is the corollary of the socialist city centre. It represents the planning ideal of creating a desired type and intensity of social interaction by physically grouping residents together in 'hygienic' conditions, and drawing them, through their propinquity, into shared day-to-day tasks. A planned systematic grouping of shops, services and facilities within such residential units is meant to rationalise residents' shopping and journey to school movements, partly to ease the transport burden and traffic congestion. Establishing services within an organically planned local area also aims to encourage residents to take a proprietary interest in their services and facilities. This aim is pursued further by the physical distinction of one neighbourhood from another, giving each an identity within overall uniformity. Such relative overall uniformity results from the pursuit of equity in access and standards, speed and economies of scale from mass construction methods.

Within urban residential planning, while the 'micro-district' has provided the main focus for communal identity and provision of day-to-day facilities, it is just one of several hierarchical levels of urban residential planning units established by socialist authorities (Table 14.2). Each level is meant to possess a stipulated population size range, relatively homogeneous flats usually of at least five storeys elevation (except in smaller urban settlements), a stipulated relationship between building area and open space, the provision of given types of service with specified service radii, and isolation from vehicular traffic. A micro-district would, for example, incorporate appropriate primary and secondary schools, convenience goods shops, dining facilities and (in the more economically advanced societies) private garaging, all adhering to a theoretical service radius of about 500 metres.

Based upon four to eight micro-districts, a 'residential district' would, at its focus, contain a major shopping complex, film theatre, library, health centre and club building, with a service area of two kilometres radius. At the level lower than a micro-district, the 'housing complex' would serve an area of radius 250 metres, providing children's play facilities and day-care centres.

With increasing levels of car ownership in Eastern European and Soviet cities,[5] however, urban residents are becoming less tied to their local area for retailing and other day-to-day activities.

TABLE 14.2
Soviet model urban residential planning: pyramid or 'layered' units

Based upon planning norms developed in 1958 and 1966, although originally promulgated in the 1935 Moscow plan.

1. Apartment house ('super-block'): 1,000-1,500 population

2. Housing complex (kvartal): 2,000-10,000 population

3. Micro-district (mikro-raion): 5,000-20,000 population

4. Residential district (zhiloi raion): 30,000-50,000 population

5. Urban district (gorodskoi raion): 100,000-300,000 population

6. Urban zone (gorodskya zona): 800,000-1 mn population

Source: Osborn (1970), 251.

Arguably, a pattern is emerging whereby in the more physically mobile socialist societies—as in developed western societies—the local residential area will hold strongest functional relevance for the very young and old. The long term planning—as well as ideological—implications of such a trend appear not to be fully appreciated in those societies. By contrast, in China, North Korea and Albania, with lower levels of economic development, private ownership of motorised forms of transport is not permitted. Ownership of bicycles does, however, appear to be near universal in these societies.

Forms of housing tenure

In an egalitarian society, urban housing should not display spatial variations in terms of quality, status, fashionability and allocatory access. Complete public control of the housing market was viewed as an essential prerequisite in overcoming such spatial variations and exploitation. In newly evolved socialist societies, private housing was usually regarded as a short term relic to be quickly eradicated. That such 'relics' may still persist is a consequence of housing shortages resulting from bureaucratic bottlenecks, higher priority for industrial investment, and changing ideological emphases. Housing construction has never been considered a productive sector, but rather a public service requiring a large (and burdensome) input of funds. After the 1949 revolution, mass housing construction in China, for example, was only possible after industrial production had been considerably increased. Especially during the Stalinist era, most socialist societies' urban authorities were compelled to deliver any financial surplusses garnered within the constraints of annual and five year plans, to the central authorities. This had deleterious effects on attempts to meet the demand for workers' housing.

Nevertheless, coupled with public land ownership and planning powers, public control of housing has aimed at giving flesh to a number of egalitarian ideals. Several dimensions may be cited:

(i) housing allocation policies which deliberately undertake social residential intermixing and the physical, if not psychological elimination of 'status' areas. Priority of access to new housing is given to young families with children, employees in key economic sectors, families living in bad, unhealthy conditions and party activists.

(ii) New urban residential construction built to a similar style, size, nature and with similar material, i.e., apartment blocks, often prefabricated. Along with requirements of economies of scale, the egalitarian aim to reduce, if not actually eliminate, spatial differentiation within the urban residential environment, has been severely constrained by the changing fashions and conventional planning wisdoms affecting density, elevation, style and other structural considerations. In Poland, for example, housing developments built in the 1950s, 1960s and 1970s, experienced successively greater population densities and increasing vertical elevation.

(iii) Living space norms attempt to equate household size with dwelling space to optimise the latter and maintain social equity. In the USSR a minimum of nine square metres per person was adopted in 1922, and households were directed to appropriately sized dwellings. By 1930, however, investment in new housing had been halved, and compulsory sharing was experienced during much of the rest of Stalin's rule, only considerably worsened by wartime destruction. Between 1950 and 1971, however, the amount of 'useful' housing space[6] in Soviet towns and cities increased by 300 per cent, and by 1977 it officially stood at 12.1 square metres per person, a little above the Eastern European norm. By contrast, Chinese living space has been considerably reduced since the 1949 revolution. In an official examination of 192 Chinese cities the figure per person had fallen from 4.5 square metres in the early 1950's to 3.6 in 1979, largely as a result of rapid urban population growth. The lower figures for China also reflect cultural conditioning, particularly in relation to the traditions of higher population densities and closer extended family networks.

(iv) Public housing rents are both heavily subsidised and directly related to a household's income - usually set at between five and ten per cent of it - and not to the nature or location of the accommodation itself. Rent considerations are, therefore, eliminated from any element of housing choice which residents might have, theoretically further reducing socio-economic differentiation between residential areas.

Only rarely, however, has socialist urban housing been totally in the hands of public authorities (with the possible exceptions of Albania and North Korea). In most of these societies. rural dwellings are held predominantly in private hands, while in the urban sector various forms of 'semi-public', 'semi-private' and private ownership are present to restrain the efficacy of ideal socialist urban residential spatial structures:

(a) 'semi-public' housing: In the 1930s, Soviety industry ministries had responsibilities for housing construction to encourage workers to their newly established factories and to reduce labour turnover. Housing and services for their employees were therefore located by factory managers close to the plant (with the consequence that such residential areas often experienced heavy industrial pollution). Local planners often played little part in such decisions, which often reflected inter-ministerial and bureaucratic competition rather than the desire to establish egalitarian housing provision. Similar activities have since been pursued in China and elsewhere. Such housing development is viewed as 'semi-public' since while it is administered by state organs, residence is usually restricted to the families of present (and perhaps past) employees of the particular enterprise. The extent of this housing and the numbers of employees involved is difficult to assess: we are told, for example, that [illegible] per cent of the residents of a particular Shanhai neighbourhood work in the local factories whose management built the area's 900 flat blocks (MacQueen, 1977).

(b) Co-operative or 'semi-private' housing: This entails groups of citizens - upto 60 in the Soviet Union—cooperatively purchasing the materials to construct, and subsequently live in, their own 'collective' housing. However, as few citizens in any socialist society could hold sufficient capital for such a venture, and especially as building materials tend to be scarce, low interest state loans are ascertained, under conditions which tend to favour higher income groups. While require to conform to general planning and housing norms, higher living space standards have been approved for some societies' co-operative housing, entailing a more lavish use of building materials than in the public sector, and thereby further depriving the latter of resources which are already in short supply. With co-operative and public housing developments being pursued concurrently, but not necessarily on the same area of the city, and with a bias towards higher income groups in co-operative housing, an implicit degree of income-based residential segregation may result.

Certainly, the more economically advanced socialist societies now tend to rely on the public housing sector only for families who are considered to be financially unable to participate in co-operative schemes.

(c) Classical Marxism regarded owner-occupation in housing: as ideologically far less significant than larger scale capitalist ownership of property. Nevertheless, Engels (1973) did view it as an inappropriate form of tenure for a mobile industrial society since it would both restrain labour mobility, and 'blunt' the rising class consciousness of the urban proletariat. Within contemporary socialist societies, the existence of private owner-occupation derives from two sources:

(i) Although landlordism has generally been abolished, the 'relic' of pre-existing owner-occupation has persisted and has latterly been enhanced. For example, the detached, wooden owner-occupied dwellings which make up 25 per cent of all Soviet urban housing can now be improved with official loans of upto 1500 roubles. Such dwellings, which were removed in the larger cities during earlier phases of redevelopment are today more important in the smaller and more remote urban centres.

(ii) Positive encouragement has been given by some authorities for flat dwellers to buy their rented accommodation. This has come about as a means of relieving public bodies of the burden of upkeep and subsidy of the flats, while it also adds to national savings, thereby helping to fund other state projects. This approach has been notable in Romania, Cuba, Poland, Hungary and more recently in

> China. In Cuba for example, the first Urban Reform Law, of 1960, enabled all existing tenants to become owners of their homes within a period of five to ten years: they would pay the equivalent of their existing rent in lieu of 'mortgage repayments'. While such property reverts to the state for re-allocation on the death of the occupants, some 85 per cent of Cuban urban households were owner occupiers by the mid-1970's.

In Romania, 'privatisation' began in 1967, when newly completed housing projects were handed over to individual private owner occupiers. During the 1960s, when the country pursued a rapid industrialisation programme, Romania invested a significantly lower proportion of its state income into housing compared to its socialist neighbours. For example, the years 1966 to 1969 saw 53 per cent of the country's fixed investment going into industry, while the figure in Hungary was 41.7 per cent and for Czechoslovakia 39.4 per cent. By contrast, Romanian housing received just 6 per cent of the funds. Consequently, with a backlog of housing shortages building up, Romania's flat purchasing policy was extended in 1974. Seeking pragmatic solutions to a worsening situation, public flat dwellers in Bucharest and other major cities, with the exception of low-income families and young couples, were given twelve months' notice to either buy their flat or submit to rent increases which would in some cases double the level of existing payments. Loans of up to 80 per cent of the purchase price at a 5 per cent interest were made available. Some spatial segregation - with a preference for purchasing the newer, outer flats, most likely to be inhabited by younger couples - was an expected although not verifiable outcome.

In China, particularly since 1980, sitting tenants have been allowed to purchase their dwelling, with payments spread over fifteen years. The state also provides aid, such as building materials, for people building their own homes. However, landlordism, while anathema to socialist, and particularly Chinese ideology, still exists in China, as in a number of other socialist states, albeit on a small scale. In the People's Republic, a family owning a home but moving away from it for reasons of temporary employment, are able to let it out for the duration of their absence. Also in China it has been officially recognised that a person owning more than one dwelling may, under certain conditions, rent out the second until the local housing bureau is in a position to purchase it at its cash value. In all cases of private renting in China, payment levels are fixed by the state to cover the costs of maintenance and repairs, but not to allow a profit for the landlord. In most of the countries mentioned, public encouragement for owner-occupation is likely to grow as the state shifts some of its financial and social responsibilities back to the family.

Within such perspectives of increasing pragmatism and a dynamism which is likely to increase measures of residential differentiation, this paper now briefly attempts to summarise the various ecological patterns which have been observed in socialist societies, in an attempt to reveal the nature of the emerging residential structures of socialist cities.

Emerging ecological patterns

In one of the earliest ecological studies of a socialist city published in English, Musil (1968) discussed the spatial structure of pre- and post-war Prague. The pre-socialist city exhibited elements of differentiation similar to those emphasized in models of western urban ecological patterns. For twenty years after the war, during the period of strongest ideological dogma, a stated aim was the elimination of major areal residential differences in age structure, marital status, fertility, birth rates, death rates and infant mortality. This was initially pursued through state dominance of the housing market and planning developments, controlling, on the one hand, access to, and standards in new housing, and, on the other, migration to and from the city.

While overall differentiation was seen to be considerably reduced, old and middle-aged residents tended to remain in the oldest (implicitly pre-socialist) housing, such that by the mid-1960s, spatial correlations could be observed between, firstly, age of housing and age of occupants, and secondly, period of development and size of household—the older dwellings generally housing the smaller households.

While the application of factorial ecology methodology to socialist cities has been limited, not least due to problems of data availability, Weclawowicz (1975, 1977, 1979) was able to use detailed 1970 Polish census data (comparing them with 1931 material) to examine urban social areas in Warsaw.[8] The principal factor in this analysis was referred to as 'socio-professional position' ("similar to ..... socio-economic status"). Positively correlated variables having high loadings on this factor included population in intellectual occupations, populations with a higher education, and the proportion of the population in the 15-64 age group. Representing the adult, educated intelligentsia, this factor had a high positive loading with publicly subsidised state housing, particularly in Warsaw's central residential areas, which were rapidly constructed in the early post-war period to meet the needs of a burgeoning bureaucracy. By 1970 such groups occupied a manifestly privileged location. They were also younger and ethnically more homogenous than the 1931 population, although 'less integrated'. Overall, a 'sector-zone-tract system' was recognised for the city, representing a spatially expressed socio-economic mosaic. In other words, urban residential segregation based upon socio-economic status, while discernible through spatial analysis, was certainly not as great as, nor derived from the same processes as those operating in 'western' cities (Table 14.3). Such as analysis, does, however, expose imbalances in accessibility, not only in respect of housing, but to other urban services and facilities. As one might expect in a planned society, spatial differentiation of land was greater in 1970 than in 1931, with functional 'wedge-shaped belts' being strongly modified by many elements of concentric and multinuclear development.

In Prague and Sofia, Carter (1979) also found a greater functional segregation, producing more regular and simpler land use patterns than in pre-socialist times.

In subsequent work, Weclawowicz (1981) further applied a principal components analysis to 1970 census data for nine other Polish cities. With minor variations, the findings from this analysis largely reinforced the themes of the Warsaw analysis: areas of relatively high socio-occupational position were found near to the city centres, whilst those of lower position were located in the suburbs. Some positive relationship was also found between age structure, household size and age of dwelling: old age, small households being particularly characteristic of the oldest dwellings, as found in Prague. From his findings, Weclawowicz felt justified in producing a table (Table 14.4) distinguishing feudal, capitalist and socialist processes which he considered to be important in shaping urban socio-spatial forms.

Overall, greater functional differentiation, coupled with an (attempted) elimination of socio-spatial segregation within urban residential areas, provide two of the most striking evolving patterns of socialist cities. However, other major dimensions of urban spatial structure may well reflect local/regional/national conditions and traditions, to evolve differing individual patterns, inspired by cultural and historical factors rather than revealing a common socialist stamp. For example, Carter (1979) has recognised an emphasis on polycentric development in Sofia, where planners tend to follow Soviet trends, whilst in Prague, where greater awareness of western trends was evident, 'wedge' or sectoral planning was gaining dominance.

TABLE 14.3

Factors affecting social segregation in socialist cities

	Measures attempting to eliminate segregation		Sources of Segregation under socialism
1a	Relative physical homogeneity of housing within between urban areas	1.	Temporal changes in conventions, wisdoms, and fashions, costs, availability of materials, labour etc. evolving different densities, building styles (eg. elevations in residential areas of different ages.
1b	Standard living space norms		
2.	Equitable rents proportional to household incomes	2.	Through flat swapping and ex-gratia payments, higher income groups able to buy their way into more desireable public housing areas (eg. late pre-socialist flats with greater living space.)
3.	State allocation according to housing need.	3.	Housing 'markets' modified by growth of cooperative and private housing reducing the relative efficacy of public housing policies.
		4.	Cooperative housing usually representing residential development for higher income groups.
		5.	Restricted journeys to work may tie one residential area to one specific industrial/ service employer.
		6.	Persistence of owner occupation in order inner-urban pre-socialist housing usually representing old age residential structure.
		7.	Private purchase of state flats liable to be concentrated in newer and better provided areas.

TABLE 14.4

'Intra-urban pattern in cities of different socio-economic formations'

Socio-economic formation	Predominant form of socio-spatial patterns in the city	Processes and forces of the mode of production shaping social pattern in the city
Feudalism	Horizontal segregation according to occupation, ethnicity and religion. Vertical Segregation according to social position.	Production structure Level of technology Guild system
Capitalism	Spatial (horizontal) segregation of social groups according to class, ethnic and religious origin (zonal, sectoral, multiple nuclei models).	Power structure Social status Rent of land
Socialism	Decline of socio-spatial segregation. Emergence of mosaic patterns. Differentiation of occupational strata.	Social value of labour force Housing policy

Source : Weclawowicz (1981), 182.

Hamilton (1979) suggested that up to eight distinctive zones could be found in European socialist cities to yield a concentric pattern of successive building stages superimposed upon a more sectoral arrangement of functional zones related to particular location characteristics, historic factors or transport functions. He classified these zones as:

(i) The historic core, wherein the city's oldest buildings may well have been restored or reconstructed;
(ii) The relict capitalist city or pre-socialist inner area, containing remnants of the commercial, industrial and residential intermixture of land uses characteristic of ad hoc growth and low transport technology;
(iii) The zone of socialist renewal, wherein such functional heterogeneity is being replaced by functionally separated uses with an emphasis upon residential development;
(iv) 1950s socialist housing;
(v) Integrated socialist neighbourhoods of the 1960s and 1970s;
(vi) Isolation belts' of green wedges acting as buffers between industrial zones and residential areas;
(vii) industrial zones;
(viii) Oopen countryside, forest or hills, including tourist complexes.

Such a classification does, however, fail to acknowledge some of the more dynamic aspects of socialist urban devlopment - in particular the apparent reverse process of 'commercialising' city centres and the increasing proportion of new housing devoted to the co-operative and private sectors. The present writer therefore suggests a simpler, yet more dynamic model (Fig. 14.1) for consideration. Admittedly, with the diverse range of cultures that have been encompassed by state socialism, any model purporting to represent trends in all such societies needs to be viewed with some caution. Even today, for example, relatively little comparative material is available on the internal nature of Chinese, Albanian and North Korena cities. What needs to be emphasised, however, is the dynamic nature not only of socialist cities and their residential areas, but of socialism itself, and the re-interpretations of it perpetrated by the various state socialist societies within their own national frameworks. Heightening economic problems are increasing the need to find pragmatic and efficient solutions rather than necessarily egalitarian answers to questions raised by new pressures. How such processes and new emphases impinge upon the spatial patterns of socialist urban development will provide no small measure of interest.

FIG. 14.1

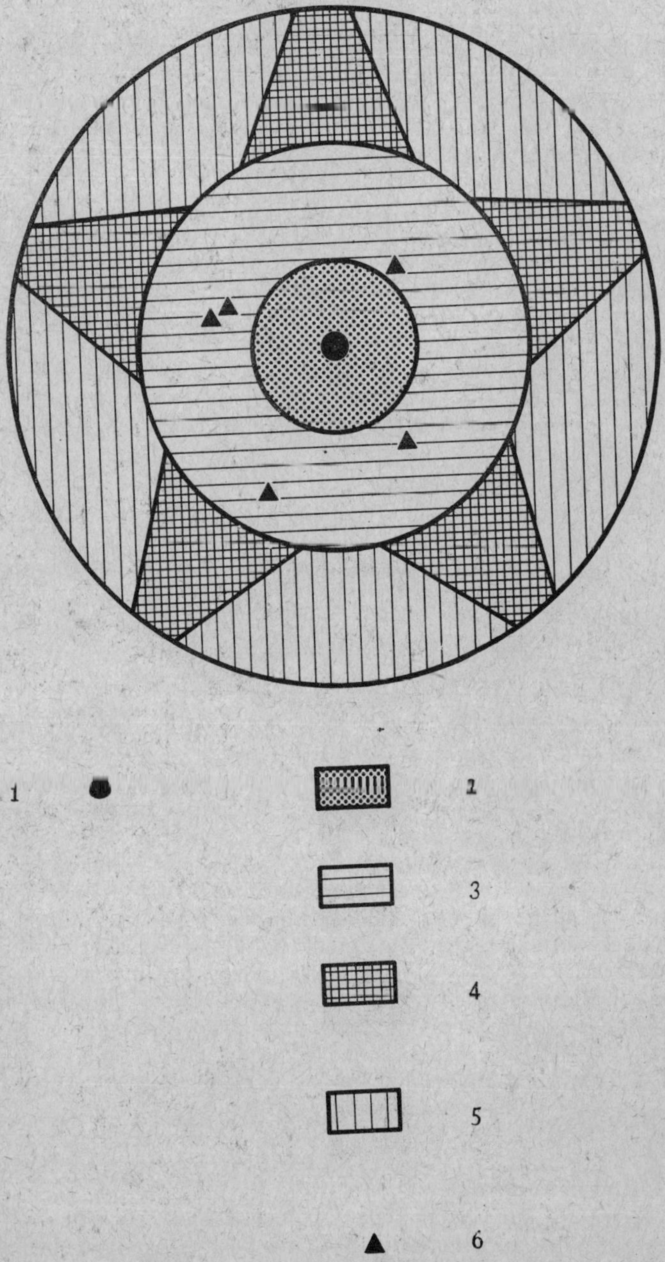

Fig. 14.1 Emerging residential structures of the socialist city: a hypothetical model

KEY to Fig. 14.

1. CCD - in some cases partially reverting to a CBD.
2. Pre-socialist inner urban: old household structure and varying degrees of owner occupation.
3. State socialist housing: relatively high income, mature household groups close to the centre, with relatively lower income groups beyond.
4. Co-operative housing: developing in extent and importance since the 1960s; relatively high income groups and relatively younger mature households. Some private ownership may also exist in these areas.
5. State socialist housing: low income groups, relatively younger,mature household unable to buy themselves into the cooperative sector.
6. Industrial housing: built specifically for a local industry's employees (usually a 'key' industry) often irrespective of local planning controls. Not found in all societies, and generally a feature of the earlier socialist phase. Segregation according to employment implicit in this type.

FOOTNOTES

1. Post-World War II for the Soviet Union and Eastern Europe (1945+), post-Korean War for North (Democratic People's Republic of Korea (1953+), and post-Vietnam War for a reunited (Socialist Republic of) Vietnam (1975+). The (1949) revolution in China followed twelve years of war - eight years of the Sino-Japanese conflict followed by four years of civil strife.
2. eg. Baranov and Davidovich in the early 1960s (Osborn, 1970).
3. A similar device had been effected in Czarist times.
4. This well illustrates the so-called 'equity-efficiency trade off', where equity appears to be traded off for the sake of efficiency.
5. Budapest, for example, has 100 cars for every 1000 inhabitants, a level planned to rise to 320 by 1990. The rate for Hungary as a whole is 50 (Moseley and Smith, 1979).
6. Including all rooms and service spaces such as kitchens and corridors in flats and hostels; this is, therefore, an inflated indicator for actual living space.
7. **Renmin Ribao,** 6 October 1979, p. 3.
8. Jagielski (1978) also used a factorial ecology framework in his analysis of Wroclaw. Deserving a wider auidence, this work has yet to be translated into English.

REFERENCES

Frank W., Carter "Prague and Sofia: an analysis of their changing internal city structure", in **The socialist city**, ed. R.A. French and F.E.I. Hamilton (Chichester: John Wiley, (1979) pp. 425-459.

Friedrich Engels, **The conditions of the working class in England**, London: Panther, (1973).

Grzegorz Weclawowicz, **Struktura przestrzeni spoleczno-gospodarezej Warszawy w latach 1931 i 1970 w swietle analizy czynnikowej** Warsaw: PAN (1975).

___ "The structure of the socio-economic space in Warsaw in 1931 and 1970", **Geographia Polonica**, 37 (1977), 201-224.

___ "The structure of socio-economic space in Warsaw 1931 and 1970: a study in factorial ecology", in **The socialist city**, ed. French R.A., and Hamilton F.E.I., (Chichester: John Wilcy, 19790, pp. 387-423.

___ "Towards a theory of intra-urban structures of Polish cities" **Geographia Polonica**, 44 (1981), 179-200.

Ian Hamilton, F.E., "Spatial structure in East European cities", in **The socialist city**, ed. French R.A. and Hamilton F.E.I., Chichseter: John Wiley (1979) pp. 195-261.

Jagielski, A.,"Struktura spoleczno-ekologiczna miast polskich a koncepcja szkoly chicagowskiej", in **Procesy urbanizacji kraju w okresie XXX-lecia** PRL, ed. J. Turowski Wroclaw: Ossolineum (1978).

James H. Bater, **The Soviet city**, London: Edward Arnold, (1980).

Jiri, Musil, "The development of Prague's ecological structure", in **Readings in urban sociology**, ed. R.E. Pahl, Oxford: Pergamon (1968), pp. 232-259.

___ **Urbanization in socialist countries**, London: Croom Helm (1981).

Moseley, M.J. and Smith, P.R., "Hungary: urban and rural planning", **Town and Country Planning**, 48, No. 5 (1979), 85-87.

Papagjorgi, H., "Peopling the countryside", **Albania Today**, 63 No. 2 (1982), 14-19.

Robert, J. Osborn, **Soviet social policies** Homewood, Illinois: Dorsey, 1970).

The World Bank, **World Development Report 1980** (Oxford: Oxford University Press (1980).

___ **World Development Report 1981** Oxford: Oxford University Press, (1981).

FIFTEEN

Ashok K. Dutt and Gerardine D'Sa

Density Gradients of Calcutta City 1872-1981

The spatial distribution of population densities has been described by several empirical functions starting with the Clark model (1951), and the subsequent models by Brush (1968) and Newling (1969)[1]. The spatial structure of differential growth is an appropriate focus of attention for geographers. Growth and decay are, after all, the measurable expressions of the processes by which the observable spatial structure of any phenomenon at a given point in time, is achieved.

Clark pointed out that the density of population was highest at the centre of the city and that it declined outwards: distance-decay is in operation. Though he based his empirical studies on several cities of Europe and North America, his data was pre-World War II, when the office and concentrated commercial function was not too well established in most western city centres. Moreover, though some of his example cities showed a clear lower density at the centres, he overlooked them and based his conclusions on straight-line fit (regression line). Nonetheless, Clark gave us a pioneering idea to statistically analyse the city density patterns with reference to a single focus.

Newling, almost two decades after Clark's findings, pointed out that the density patterns in the cities are dynamic and respond to the age of the city. He pointed out that at the youthful stage a city adheres to Clark's distance-decay model, but as the city becomes older the commercial function dominates the centre in such a fashion that it undergoes depopulation. Thus, a crater-effect is found in the density gradient with a low density central area, its immediate rim with the highest density and thereafter, following a distance-decay pattern outwards. Though Newling tried to design his model for universal application, it can only be applied to United States' cities mainly because the cities there do not have a history before the 17th century.

Brush, in an innovative research, applied the density gradient idea to Indian cities and found out that the traditional cities demonstrated Clark's distance-decay paradigm as the central bazaars housed a considerable number of residents. However, in the colonial cities (Calcutta, Bombay and Madras), established as a result of British rule, he discovered the Newling type of crater effect at the centre.

The purpose of this paper is to study Calcutta's density gradient historically and find out whether Newling's youthful stage can be identified from the earlier spatial pattern.

Methodology

The first step in the population density analysis of a city, beyond obtaining the areal and population statistics for municipal wards, is finding the nodal point to which the spatial structure of population can be related. Distances are measured between the areal centres of each ward and the city nodal centre. Such a calculation also assumes that areas at the same distance from the centre, also posses similar morphology. In Calcutta, except in the extreme West and South, there are no discernible changes in its morophology.

The difficulties encountered in this study are that population density figures are based on the measurement of gross area of the entire ward and do not represent net residential areas. Population totals in this study also include migrant workers who stay at their workplace for the week and return to their rural home for their weekly holidays.

In any city, the changing spread of population density over time is an indicator of its growth and urbanisation. This paper assesses the spatial patterns of population density in Calcutta, and analyses these patterns in the context of its historical roots, socio-economic status and technical association, during the past 110 years, (i.e., from the first Census in 1872) through pre-and-post-independence times.

The population figures for 1872, 1891, 1911, 1931, 1951, 1971 and 1981, have been analysed spatially and temporally, using SYMAPS and APL Polyfit curves. The Bipin Bihari Das Bagh (former Dalhousie square and also used to be known as

Lal Dighi) has been taken as the nodal point, since here communication is greatest and land values highest. It is at the heart of the only office-cum-commercial (non-residential) area of the city. As the city limits were restricted with a short distance from the CBD at the earlier census years the curve lines extended accordingly for those years.

Historical background

The origins of Calcutta, according to British sources, date back to 1690 when it grew to be a significant possession of the British East India Company. However, its actual origin as an urban trading outpost may be assigned as at least 50-60 years earlier when the Portuguese and Dutch established outposts there and trading flourished out of the Sutanuti market, one of the three nuclei of Old Calcutta. It was the capital of British India from 1772 to 1911. As a creation of British colonialism this presidency town had, and still enjoys, a Western-style low population density Central Business District (CBD), situated around the original nucleus - Fort William - with banks, insurance offices, cinema halls, hotels and large-scale retail establishments at the city centre[2]. Another centrally located area has developed, adjacent to the commercial core, with the court-houses, Government offices, Railway Headquarters, Customs Buildings, and General Post Office. These areas have had little residential population. To the North of the CBD, is the traditional central bazaar (known as Bara Bazaar), combining high density commercial and residential uses, densities reaching 1100 persons per acre at places. The former European Settlement was about 5 km. to the south of the CBD and connected to it by Chaurangi Road. This road is now one of the main activity centres of the metropolis and population densities along its western margin have increased considerably during the last century.

In 1947 there was a sudden influx of people, mainly those displaced by the partition of the country, in search of work. Since the country's independence, in 1947, there has been a significant decrease in the European population of Calcutta, accompanied by an increase in Indian population and a rapid expansion of the city limits, to keep pace with increasing population concentration within it.

Spatial Distribution of Population in Calcutta: 1972-1981

The areal spread of population concentration in Calcutta for the years 1872, 1891, 1911, 1931, 1951, 1971 and 1981, has been studied through a series of maps. These isopleth maps show lowest densities in the Central Business District, and highest densities just northwest of it—the traditional central bazaar (Bara Bazaar) of high residential and commercial activity. These two areas represent the old city component, and have served as the core of Urban Development.[3]

The maps show that the difference in highest and lowest densities has increased with time. In 1872, half the population lived within two kilometres of the city centre. In 1891 the radius of mean population spread widened to two and a half kilometers and three km in 1911 and 1931. In 1951 population densities in north Calcutta increased considerably reaching a peak in 1971—mainly due to the influx of workers in search of employment in the city. These people lived in crowded tenements close to their place of work to avoid the cost of using public transport. Since 1981 there has been a slight decrease in highest densities, resulting from an outward spread of population, following recent increases in more economical and efficient public transport networks.

In the case of Calcutta, the most obvious influence on population density has been the age and mode of development and building - the oldest areas having the highest densities. The lowest densities continue to be in the lowlying areas of the South and West - the newly developing parts of the city.

The most desirable residential location has been centrality or maximum accessibility near the city centre, since transport routes and commercial activities converge there. In contrast to accessibility the higher rent functions near the city centre tend to limit residential landuse there. Thus residential landuse is a function of both, land capability, and accessibility[4].

In Calcutta, the poorer people, in general, live close to the centre on rather expensive land but using very little of it, the central residential blocks being very small and crowded together. Thus the smallness of the homes (mostly one-room households) reduces their cost. Besides transport costs are eliminated because these areas are within walking distance of the place of work. The richer people usually live away from the crowded commercial areas of the city centre in larger houses. Since land consumed by each household increases with distance from the city centre, population densities must drop with due allowance for the variation in the size of the household[5].

In the last decade the area of densest population has shown a marginal decrease, as the mass of working men have been moving to the outlying districts. This is probably the result of rehousing schemes, the redistribution of working places, and concessions on long-distance fares together with increased frequency in public transportation[6].

Calcutta developed a western-style CBD true to its characteristics of a colonial city, with a high concentration of mercantile-type office functions, retail trade, and a low density of residential houses. The administrative quarters consisted of the viceroy's (later Governer) house, the main government offices, the court houses and the general post offices. In the CBD area also there were Western-style hotels, churches, banks, and museums as well as occasional statues of British sovereigns, viceroy and other dignitaries at important road intersections, reminiscent of royal eternity[7].

Density Gradient Changes in Calcutta 1872 - 1981

The development of the city through the interaction of traditional and colonial factors, the incorporation of neighbouring settlement modes, and the rapid outward expansion of the city perimeter, have produced a highly complicated spatial structure of population density gradient that cannot be adequately described in terms of simple linear distance from the city centre[8]. However, this study for comparative temporal studies, analyses them using polynomic lines for a clearer explanation of the pattern.

The population density gradient curves have been prepared for the years 1872, 1891, 1911, 1931, 1951, 1971 and 1981 (Figure 15.1) using the APL Graphics (Library NO. 23) of the Akron University Polyfit program. In general, all the seven polynomial curves show a low density at the city centre—the Central Business District—with density rising to a peak about a kilometre away from the centre, and declining progressively towards the periphery. The gradient of the curve is gradual in 1872 and steepest in 1971, all exhibiting the crater effect but in varying degrees of slope.

A closer examination of the density gradient curves reveals that from 1872 to 1951 both the central low density, and the bordering peak density, showed increases and gradients, increased only marginally so the curves have a similar trend. From 1951 onwards there is a decrease in central density, together with a rapid increase in the peak densities of the crater rim. Also until 1971, there exists a subsidiary peak of population density, moving progressively away from the city centre, and representing nodes of residential concentration away from the CBD.

The striking feature brought out with the superimposition of these population density distance decay curves, is the intersection of the 1951, 1971 and 1981 curves about half a kilometre away from the city centre, and again at four, and six kilometres away from the city centre (Figure 15.1). It shows how commercial landuse is displacing residential landuse on the valuable centrally located land. Besides in post-independence times, there has been an outward shift of the population—bridging the gap between neighbouring residential areas. The 1981 population density peak is lower than that of 1971, while the former European residential areas are now occupied by Indians and increasingly used for office purposes. Thus, there is also a gradual decline of the secondary peak from 1951 to 1981.

FIG. 15.1
CALCUTTA CITY

Density Gradient Curves (APL Polyfit) for the years 1872, 1891, 1911, 1931, 1951, 1971 and 1981

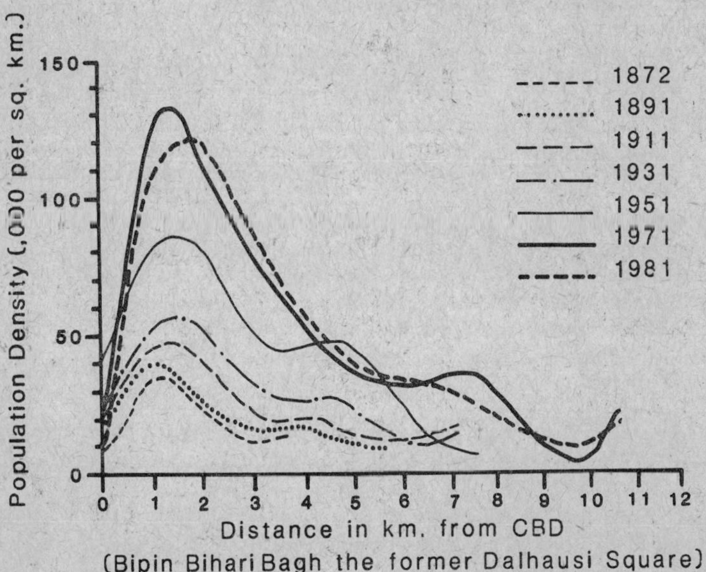

SOURCE: Based on Census Data

From this study of population density gradients in Calcutta it is evident that the city centre has, throughout this span of 110 years, been depopulating. Population rises to the rim of the crater and then progressively declines. The secondary peak about 5 km away, representing the residential areas of the richer classes, was pronounced till 1951 and has since declined, indicating the general spread of population and the lack of development of separate residential areas in post-independence times.

Conclusion

The period before 1872, unfortunately, could not be analysed due to the lack of population statistics. This 110 year study covers two stages of development—the pre-independence colonial city, and the post-independence mature city. Basically the pre-independence pattern has been retained after independence as the centre continues to be dominated by commercial and government related activities. Moreover, increasing commercial and government activities after independence has exaggerated the CBD's crater effect in the population density distance-decay curve, and stretched the city's outer limits beyond former residential suburbs, to merge into a larger conurbation. The trend towards central depopulation and the spread of population in the peripheral areas is a clear indication of rapid urbanization in years to come.

At least since 1872 the central area of Calcutta showed depopulation reflecting Newling's "mature and old stages", but the historical records prior to 1872 show that the centre housed European residents in limited numbers at least until the beginning of the 19th century with a relatively lower density compared to the natives who lived in their thousands in densely packed areas around the European nucleus. Non-residential function of the central area was greatly accentuated after railways were introduced during eighteen fifties, boosting mercantile function of the central area. If the pre-1872 data was available possibly a low density at the centre may be seen starting from Calcutta's initial days which then would challenge Newling's formulation of having higher density at the centre during the youthful stage of the city. A city with a Colonial base, like that of Calcutta, possibly always had a low central density compared to its immediate ring.

1. Brush, John E., "Spatial Pattern of Population in Indian Cities", **Geographical Review**, 58 (3), pp. 362-391; Clark Colin, 1951, "Urban Population Densities", **Journal of the Royal Statistical Society**, Ser. A, CXIV (4), pp. 490-496; and Newling, B.E., 1969, "Spatial Variation of Urban Population Densities", **Geographical Review**, 59 (2), American Geographical Society, New York,(1968), pp. 440-442.

2. Brush, **op. cit.**

3. Breese, Gerald, "Urban Development Problems in India" **Annals of the Association of American Geographers,** 52 (2) (1963), pp. 253-265.

4. Newling, B.E., "Urban Growth and Spatial Structure: Mathematical Models and Empirical Evidence", **Geographical Review,** 56 (2) (1966), pp. 213-225.

5. Berry, Brain J.L., Simmons, J.W. and Tennant, R J., "Urban Population Densities", **Geographical Review,** 1963 Vol. LIII, No. 3, American Geographical Society, New York, pp. 389-405.

6. Simeon, D., "The Rate of Change of Population Densities in the Concentric Zones of a Metropolis", **Ekistics,**30 (1976), Athens, (1970), pp. 8-13.

7. Dutt, Ashok K., "Cities of South Asia", in Stanley Brunn and Jack Williams (eds.), **Cities of the World,** Harper and Row, New York, (1983) pp.337-338.

8. Papageorgiou, G.H., "A Generalization of the Population Density Gradient Concept", **Geographical Analysis,** 3 (2), Columbus, (1971), pp. 121-127.

SIXTEEN

Ian McAllister and Jonathan Kelley

Contextual Characteristics of Australian Federal Electorates

It has long been evident that the context in which people live effects how they live. For example, living in high status communities rather than in low status communities influences voting behavious, social and political attitudes, economic status, and the nature of social networks (Cox 1972; Fitten, 1973; Linz, 1968; Parcel, 1979; Segal and Meyer, 1969; Sewell and Armer, 1966; Wright, 1977). These effects are best understood in the case of voting, where strong circumstantial evidence suggests that when one social class is dominant in a constituency, a larger proportion of the electorate votes in accordance with that class's predisposition (Butler and Stokes, 1974:140-151; Miller, 1978: 50-51). In an influencial article, Putnam (1966) has argued that differences in social contacts explain this effect, while Przeworski and Soares (1971) have refined the simple relationship by suggesting that, in certain circumstances, high concentrations of one class may initiate a reactive increase in class consciousness within the other class.

But contextual effects do not stem only from socio-economic status. The ethnic composition of a community affects attitudes to racial and ethnic questions, individual modernity, earnings of both majority and minority groups, and income inequality (Frisbie and Neidert, 1977; Nachmias, 1979; Studlar, 1977). Other community characteristics have a wide variety of influences, for example, on political behaviour, income, migration, modernisation, and the propensity to strike (Firebaugh, 1979; Lincoln, 1978; Sly and Tayman, 1977; Timms, 1971). These contextual effects can sometimes be highly significant. For example, after an extensive quantitative analysis, Miller (1978: 65) concluded that in Britain "the class characteristics of the social environment have more effect on constituency partisanship than class differences themselves, [and] perhaps much more."

Research on contextual effects in Austraila has, however, been severely limited, mainly by the lack of appropriate ecological data. There are excellent studies on the social ecology of neighbourhoods in cities (Burnley, 1972; Jones, 1967, 1969; Logan et al., 1975; Sweetser, 1978; on New Zealand, see Bowman and Hoskins, 1971 and Trlin, 1975), but none for the nation as a whole. Only one paper has been prepared utilising data for parliamentary constituencies and that simply ranks the constituencies by nine selected census variables (Legislative Research Service, 1980). Constituencies are the natural ecological unit for political analysis, since electoral outcomes are ultimately determined by how the constituencies rather than the individual electors voted, and in Australia they are also the natural unit for most other questions since Australian surveys typically sample by constituency, which allows for the identification of respondents who share a common environment at that level. Census data aggregated to Federal Electoral Divisions were first made available for the 1971 census (ABS, 1971) and again for the 1976 census. But unlike in Britain, where the introduction of similar data in 1966 stimulated a plethora of research (Barnett, 1973; Crewe and Payne, 1972; Rasmussen, 1973), Australian social scientists have ignored their potential. This paper seeks to remedy that neglect.

The aims of the paper are twofold. Firstly, to distill the wide range of census variables into a small number of conceptually and statistically unambiguous scales, each consisting of a number of individual census variables.[1] For example, socioeconomic status is measured not by the usual crude manual/nonmanual dichotomy, but more reliably and comprehensively by taking account of detailed information on occupation, education, income, and home ownership. The use of multiple indicators for each concept avoids the very real possibility of being misled by the idiosyncratic features of any one indicator.[2] Secondly, the paper aims to provide data for each of these scales, enabling other researchers to simply and routinely incorporate this information into existing surveys. We expect these data to be of considerable use to other researchers. They will be of obvious use for researchers interested in studying any of a wide range of contextual effects. They will also be of use to researchers working on aggregate level analyses—flourishing traditions in political science and history, as well as in sociology.[3] And they shall be of use as control variables for some researchers studying the effect of individual level variables, particularly in the areas of education, occupation, income, ethnicity and rural/urban differences. The reason for this last is that, in some circumstances, the omission of contextual variables leads to biased estimates of the effect of individual level variables so researchers will want to include the appropriate contextual variables in the analysis even though contextual influences themselves are not of interest.[4]

DATA AND METHODS

The basic data are from the 1976 Australian census aggregated to Federal Electoral Divisions and provided on computer tape by the Australian Bureau of Statistics (see ABS, 1978 for detailed description and codes). Given the large number of cases in each of the constituencies—a mean of 109,051 per constituency with a minimum of 76,185 in Wilmot, Tasmania—sampling error is not a problem.[5] The analysis is based on the 124 constituencies existing at the November 1977 boundary redistribution. The variables used in the analysis are described in more detail below.

The study is based on factor analysis, a technique which has dominated the field of factorial ecology since the pioneering work of Shevky and Bell (1955; see also Janson, 1969, 1980; Jones, 1967, 1969; Selvin and Hagstrom, 1963; Sweetser, 1969, 1978). Given the correlations among a large set of variables (here, census variables describing parliamentary constituencies), factor analysis obtains mathematically optimal estimates of a simple measurement model in which the observed variables are taken as imperfect indicators for one (or more) underlying hypothetical traits and the correlations among observed variables explained solely on that basis. This simple measurement model is precisely what is intuitively suggested when it is claimed that, for example, the proportion of a constituency's population with high ranking occupations, university education, and high incomes are all indicators of the socio-economic status of the constituency. If the model is true, factor analysis provides mathematically optimal estimates of the underlying traits; if the model is only approximately true, in practice factor analysis has been consistently shown to provide robust and appropriate estimates in general and for ecological analyses in particular.[6] Detailed technical descriptions are readily available for the general procedure (Harman, 1976; Lawley and Maxwell, 1971) and for its applications specifically to ecological variables (see the reference just cited); for a particularly lucid nontechnical discussion, see Kim and Mueller (1978).

The results presented here are based on a principal components factor analysis with unities in the main diagonal; factors with an eigenvalue less than one were omitted and the remainder (if more than one) rotated to simple structure by the varimax criterion. This is an appropriate and conventional choice of technical options but other reasonable choices would lead to virtually identical results (Janson, 1980; 439-441).

We use the results of our factor analysis to construct multiple item scales to measure the underlying concepts. This is preferable to simply choosing a single variable to represent the concept for several reasons. First, there will be more random measurement error in any single item than in a multiple item scale and, since measurement error leads to bias, it is clearly preferable to minimize it (Bohrnstedt and Carter 1971:130-140). Second, there is always a small but genuine risk that the single item chosen is atypical in a way not revealed by the factor analysis and this is less damaging if the atypical item is only one of many in the scale. For these and related reasons, multiple item scales are generally preferred in other contexts (for example, in attitude research and psychological measurement) and the arguments apply equally to the context at hand, although in practice the choice may not make much difference. There is less consensus about precisely how to construct multiple item scales but in practice it rarely makes much practical difference (see Kim and Mueller 1978:69-72 and the studies cited there). Our procedures are described at appropriate points in the text below.

RESULTS

Initially, a wide range of variables were included in the analysis, covering age, occupation, residence, birthplace, parents' birthplace, religion, education, income, housing, and general amenities.[7] From these numerous but often overlapping variables, we extracted four major clusters on the basis of a series of preliminary factor analyses.[8] The final result was four clusters of variables reflecting the major dimensions of social structure in Australian parliamentary constituencies: ethnicity, socio-economic status, familism, and urban-rural differences. We take up each in turn.

Ethnicity

Ethnic factors have emerged consistently in ecological analyses in Australia and throughout the world (Janson, 1980: 447; Jones, 1969: 66-86; Sweetser, 1969: 433-444). In preliminary analyses, we considered a wide variety of data on ethnicity, which fell into two main categories. First, there were measures of the proportion of the population in each constituency born in various foreign countries. The largest of these were those born in the British Isles (a mean of 8.0 per cent of the population in each constituency), followed by those of Italian (1.9 per cent) and Greek (1.0 per cent) origin, with no other group as large as 1 per cent. The main analysis is based on these variables. Second, there were a variety of measures which made no distinction by country of origin but instead essentially reflected the presence or absence of immigrants from all countries together (or sometimes all non-English speaking countries together). These included parents' birthplace, language, age and length of residence in Australia. On the basis of a lengthy series of preliminary analyses, we eventually decided not to use these variables. The basic difficulty was that they blurred the sharp distinction between the ethnic groups (described below), without adding anything new (since they essentially reflected the total number of ethnics of all groups combined); such information as these variables contained could therefore better be extracted from the first group of variables.[9]

The final analysis is therefore based on variables reflecting the national origins of immigrants.[10] From the data in Table 16.1, it is clear that immigrants are not all concentrated in the same constituencies, on the contrary, there are clear differences between various immigrant groups.

For example, constituencies that have a relatively high proportion of immigrants from the British Isles are by and large not the same ones that have a high proportion of Italian (r=.02) or Greek (r= -.10) immigrants. It is therefore misleading to think simply of 'ethnic' contexts in Australia; rather, one must think of several separate types of ethnic contexts (see Burnley, 1974: 165-183). The factor analysis in Table 16.2 shows that there are basically three types. First, there are constituencies with high proportions of immigrants from the Mediterranean countries, mostly from Italy, Greece, and Yugoslavia with some from Turkey, Lebanon, Malta, Egypt and Cyprus. Second, there are constituencies with relatively high proportions of immigrants from Eastern Europe, mostly from Poland and Austria, but also from Hungary, Czechoslovakia and the USSR. These are not in general the same constituences with high proportions of Mediterranean immigrants. Finally, there are constituencies with large numbers of Northern European immigrants, largely the British Isles. These are clearly differentiated from the constituencies with large numbers of Mediterranean or Estern European immigrants. In short, patterns of residence indicate that Australia has three distinct types of immigrant communities: Mediterranean, Eastern European, and Northern European.[11]

TABLE 16.1
Ecological Correlations Between Major Birthplace Groups [a]

Birthplace	1	2	3	4	5	6	7	8	9	10	11	12	13	14	15
Mediterranean															
1. Italy															
2. Greece	.60														
3. Yugoslavia	.53	.52													
4. Turkey/Lebanon	.40	.61	.43												
5. Malta	.37	.25	.55	.23											
6. Egypt	.43	.73	.44	.63	.37										
7. Cyprus	.63	.83	.66	.60	.57	.79									
Eastern Europe															
8. Poland	.26	.41	.41	.19	.32	.27	.40								
9. Austria	.02	.16	.10	.01	.06	.16	.15	.38							
10. Hungary	.14	.39	.18	.21	.09	.39	.36	.58	.71						
11. Czechoslovakia	.10	.36	.20	.21	.11	.42	.35	.62	.76	.95					
12. USSR	.33	.40	.36	.44	.24	.41	.46	.60	.35	.61	.63				
Northern Europe															
13. British Isles	.02	-.10	.04	-.11	.01	-.03	-.05	.09	.22	.09	.12	.04			
14. Germany	.22	.14	.40	.00	.37	.12	.23	.52	.46	.43	.44	.33	.48		
15. Netherlands	-.10	-.28	-.03	-.27	-.02	-.21	-.23	-.03	-.07	-.07	-.07	-.13	.57	.53	

Notes:
[a] N=124 Australian parliamentary constituencies.
Source: Computer tape of the 1976 census allocated by Federal Electoral Division, supplied by the Australian Bureau of Statistics

TABLE 16.2

Ethnic Residential Patterns : Factor Loadings for Major Birthplace Groups[a]

Birthplace[c]	Mean per cent in each Constituency	Factor Loadings (varimax rotation)[b]		
		Factor I	Factor II	Factor III
Mediterranean[d]				
Italy	1.93	.76	.01	.05
Greece	1.05	.78	.30	-.24
Yugoslavia	0.97	.79	.09	.21
Turkey/Lebanon	0.35	.68	.15	-.31
Malta	0.23	.65	.00	.27
Egypt	0.21	.73	.30	-.21
Cyprus	0.15	.90	.24	-.10
Eastern Europe[e]				
Poland	0.40	.35	.64	.18
Austria	0.40	-.06	.81	.18
Hungary	0.20	.13	.94	.00
Czechoslovakia	0.11	.13	.96	.01
USSR	0.10	.40	.65	.03
Northern Europe[f]				
British Isles	7.99	-.04	.11	.75
Germany	0.77	.25	.43	.77
Netherlands	0.67	-.16	.08	.84

Notes:

a N = 124 Australian federal electorates.

b Varimax rotated factor loadings from a principal components analysis with unities in the main diagonal. The eigenvalues of the first three factors are 5.80, 2.80 and 1.88 respectively; no other factor had an eigenvalue greater than 1.

c Groups not shown make up only 3.8 per cent of the total overseas born population.

d The scale used subsequently was constructed by summing the per cent in each constituency born in Italy, Greece, Yugoslavia, Turkey, Lebanon, Malta, Egypt, or Cyprus.

e The scale used subsequently was constructed by summing the per cent in each constituency born in Poland, Austria, Hungary, Czechoslovakia, or the USSR.

f The scale used subsequently was constructed by summing the per cent in each constituency born in the British Isles, Germany, or the Netherlands.

Source: As for Table 1.

The ethnic composition of each constituency was measured by three scales, one giving the proportion of the population born in Mediterranean countries, a second giving the proportion born in Eastern European countries, and the third giving the proportion born in Northern Europe. An alternative procedure for constructing scales would have been to use factor scores, but we preferred the conceptually simpler procedure; in practice the choice does not really matter.[12] Scores for these variables for each parliamentary constituency in Australia are given in the appendix.

Socio-economic status, familism and urban-rural differences

To measure socio-economic status, we began with a wide variety of variables related to the classical elements of socio-economic status: education, occupation, income, and consumption variables. Then on the basis of a series of preliminary factor analyses, the wide range of such socio-economic variables in the census was winnowed down to the seven relatively clear and logical indicators shown in Table 16.3, focussing on the central elements of education, occupation and income. In addition to these socio-economic variables we considered a wide range of other variables were included which had proven valuable in ecological analyses of other countries or which seemed useful for other reasons. These included religion, age, employment, and measures focussing on urban-rural differences and housing amenities. In the event, two additional clusters of variables (shown in Table 16.3) showed reasonably clear and coherent patterns and were included in the final analysis.

The dimensions that emerged from a factor analysis of these variables are those which have consistently been extracted from similar analyses in industrial societies throughout the world (Janson, 1980: 446-8; Jones, 1969; Sweetser, 1969; Timms, 1971). First, and most significant is socio-economic status, which distinguishes electorates where a relatively high proportion of the population has advanced education, high ranking occupations, high incomes, and high mortgage payments. (That it is the first factor does not mean it is the most important, only that there were many socio-economic variables in the analysis). These traits are strongly correlated, constituencies high in one clearly tending to be high in the others, (see Table 16.3), and show a very clear pattern in the factor analysis (see Table 16.4, first factor).

We constructed a scale measuring the socio-economic status of constituencies by conducting a second prinicpal components factor analysis including only the seven socio-economic indica-

TABLE 16.3

Ecological Correlations between Selected Socio-Economic and Other Census Variables[a]

Variables	1	2	3	4	5	6	7	8	9	10	11	12	13	14
Socic-economic Status														
1. Per cent of population with schooling to age 17 or 18														
2. Per cent of population with annual personal income over $ 18,000	.85													
3. Per cent of population with annual personal income over $ 12,000 and under $ 18,000	.78	.85												
4. Per cent of labour force in professional occupations	.89	.84	.82											
5. Per cent of labour force in administrative occupations	.73	.82	.76	.78										
6. Per cent of labour force in transport & communications	-.78	-.71	-.67	-.77	-.71									
7. Per cent of dwellings with monthly mortage payments over $ 325	.67	.75	.69	.59	.82	-.62								
Familism														
8. Percentage of dwellings with sole use of kitchen	.13	.29	.40	.30	.38	-.20	.16							
9. Per cent of dwellings with sole use of bathroom	-.22	-.09	.02	-.06	.08	.05	-.01	.45						
10. Per cent of population purchasing homes	.10	.14	.38	.28	.38	-.23	.28	.49	.49					
11. Per cent of dwellings with one or more cars	-.26	-.01	.11	-.08	.12	-.08	.07	.54	.73	.51				
Rural														
12. Per cent of labour force farmers or fishermen	-.35	-.23	-.37	-.40	-.31	.09	-.18	-.08	.17	-.45	.40			
13. Per cent of labour force employers or self-employed	-.24	-.12	-.29	-.27	-.10	-.06	-.05	.04	.23	-.35	.46	.95		
14. Per cent of dwellings with three or more cars	-.21	.00	-.06	-.15	-.06	-.07	-.06	.23	.37	-.09	.67	.82	.81	
Mean	11.3	1.0	3.0	22.6	6.4	4.9	1.0	99.4	99.3	39.2	84.5	7.5	13.7	8.4
Standard deviation:	4.5	0.9	1.5	8.8	2.6	0.9	0.6	0.2	0.5	13.8	7.4	9.9	7.3	2.9

Notes:

[1] N = 124 Australian federal electorates.
Source: As for Table 16.1.

TABLE 16.4

Factor Loadings and Factor Score Coefficients for Socio-Economic and other Variables[a]

	Factor Loadings[b] (Varimax rotation)			Factor Score Coefficients[c]		
	Factor I	Factor II	Factor III	Factor I	Factor II	Factor III
Socio-economic Status						
Percent of population with schooling to age 17 or 18	.92	-.18	-.17	.16		
Percent of population with annual personal income over $ 18,000	.94	-.02	.01	.17		
Percent of population with annual personal income over $ 12,000 and under $ 18,000	.[illegible]7	-.19	.22	.16		
Percent of labour force in professional occupations	.[illegible]0	-.21	.06	.16		
Percent of labour force in administrative occupations	.[illegible]8	-.09	.23	.16		
Percent of labour force in transport or communications occupations	-.86	-.11	-.04	-.15		
Percent of dwellings with monthly mortgage payments over $ 325	-.[illegible]1	-.02	.09	.15		
Familism						
Percent of population purchasing homes	.19	-.40	.80			.29
Percent of dwellings with sole use of kitchen	-.[illegible]7	-.04	.73			.29

se

Table 16.4 continued...

Percent of dwellings with sole use of bathroom	.11	-.19	.82		.32
Percent of dwellings with one or more cars	.02	-.46	.83		.33
Rural					
Percent of labour force farmers or fishermen	-.22	.96	.05	.36	
Percent of labour force employers or self-employed	-.07	.97	.03	.36	
Percent of dwellings with three or more cars	-.01	.88	.31	.34	

Notes:

[a] N = 124 Australian federal electorates.
[b] Varimax rotated factor loadings from a principal components factor analysis with unities in the main diagonal. The eigenvalues of the first four factors are 6.01, 3.45, 2.15 and 0.64, respectively.
[c] These are not from the varimax rotation but are based on three separate principal components factor analyses, one for each scale. Each of these included only the items in its scale and resulted in only a single factor with eigenvalue greater than 1. Scales were constructed by multiplying a constitiency's value on each variable by the variable's factor score coefficient and summing.

Source: As for Table 16.1.

tors (which showed only a single factor with an eigenvalue greater than one) and using the factor scores from that (see table 16.4). The resultant score is a weighted sum of the component items, the weights being chosen in a way that optimally predicts the underlying factor (Kim and Mueller, 1978: 60-73). This choice of scoring procedures avoids some technical problems but in practice other reasonable procedures would lead to essentially identical results.[13] The scale has no natural metric, so we re-scored it into a more intuitively meaningful metric, so that the constitueny with the highest socio-economic status (Bradfield, NSW) obtained a score of 100 and the one with the lowest status (Chifley, NSW) a score of 0.[14] This linear transformation is purely cosmetic. It has no effect on the rank or relative position of constituencies, nor on correlations, standardized regression coefficients, or the like; the only change is that metric regression coefficients are converted into a more intuitively appealing metric. Scores for each federal electorate in Australia are shown in the Appendix.

The second factor is familism, a dimension appearing in somewhat diverse forms and under different labels, in ecological studies throughout the industrial world (Janson, 1980: 446-447; Jones, 1969: Ch. 4; Sweetser, 1969: 429-433). Familism is generally interpreted as reflecting the stage in the life-cycle when the family has children and hence seeks large, unshared, often suburban housing suitable for a large family. In our data, constituencies high in familism ar marked by relatively high proportions of the population owning rather than renting their homes, with sole use of bathroom and kitchen, and with the use of a car. A scale of familism was constructed from these indicators, weighting each according to factor scores from a separate factor analysis including only familism variables. Since, the scale has no natural metric, it was transformed so that the constituency highest in familism (Diamond Valley, NSW) has a score of 100 and the lowest (Melbourne, Vic.) a score of 0.[15] Again, scores for each constituency are given in the Appendix.

The final factor is the familiar distinction between rural and urban areas. In these data it is indicated by the per cent of the electorate's labour force employed in agriculture, the per cent self-employed, and the per cent with three or more cars. We constructed scale scores for rurality by the procedures described above and converted them to a metric of 100 for the most rural constituency (Maranoa, Qld.), to 0 for the least rural (Sydney, NSW).[16] The final scores for each constituency are given in the Appendix.

Relations among the scales

These six scales measure diverse aspects of Australia's federal electorates and hence are not highly correlated. None of the correlations between scales, shown in Table 16.5 exceed .50 and most are much smaller. Note, first, the clear distinction between ethnic groups.

Electorates with relatively high proportions of Mediterranean immigrants are somewhat more likely to have disproportionate numbers of Eastern European immigrants but no more (or less) likely to have Northern European immigrants. And, perhaps surprisingly, electorates with Mediterranean or North European immigrants are neither noticeably higher or lower than average in socio-economic status while, in contrast, those with East European immigrants tend to be relatively high status. Northern European electorates differ in that they tend to be high on familism while, in contrast, electorates with other kinds of immigrants tend to be low on that dimension. All immigrant electorates tend to be more than the average. Second, socio-economic status is not closely related to other characteristics of electorates. High status electorates are, to be sure, a little more likely to be urban rather than rural, and to have relatively more Eastern European immigrants, but otherwise differences are conspicuous mainly by their absence. Nor are the electorates high on the familism index especially distinctive. In particular, almost as many of these generally suburban electorates are low in socio-economic status as are high in status—being suburban is not the same as being rich. Finally, rural constituencies are, as is well known, rather less likely to have immigrants and somewhat lower in status than urban electorates.

DISCUSSION

Traditionally, social science has postulated two sets of independent variables to account for social and political behaviour: on the one hand, there are the individual characteristics of the person, such as his social class, age, sex, and ethnic origin; on the other hand, there are the characteristics of the context in which he or she lives and works—the class milieu, the urban, suburban and rural six, the ethnic and religious composition, and so on. Observed patterns of behaviour often result from the interaction of individual and contextual characteristics. The data we have presented allow these questions to be addressed in the Australian context with a simplicity and clarity heretofore impossible.

The purpose of this paper has been to produce a small num-

TABLE 16.5
Correlations between Ecological Scales[a]

Scale	Mediterranean 1.	East Europe 2.	North Europe 3.	SES 4.	Familism 5.	Rural-Urban 6.
1. Mediterranean Ethnicity[b]						
2. Eastern European Ethnicity[b]	.38					
3. Northern European Ethnicty [b]	-.03	.17				
4. Socio-economic Status[c]	-.11	.45	.13			
5. Familism[c]	-.30	-.28	.46	.16		
6. Rural-Urban[c]	-.48	-.48	-.32	-.30	.05	
Mean	4.9	1.2	9.4	27.2	43.7	26.3
Standard Deviation	5.2	0.9	5.1	18.9	20.2	24.0

Notes:

[a]N = 124 Australian federal electorates. Scales constructed as described in the text.

[b]Scored as the percentage of the constituency's population born in the indicated countries. See Tables 16.1 and 16.2 for details.

[c]Scored in a conventional metric such that the constituency lowest on the variable has a score of 0 and the constituency highest on it has a score of 100. High scores indicate constituencies with, respectively, high levels of socio-economic status, many family type suburban dwellings, and more farmers and fishermen.

Source: Data shown in the Appendix, based ultimately on data from the 1976 census allocated by Federal Electoral Division, supplied by the Australian Bureau of Statistics.

ber of conceptually clear scales to represent the contextual charactheristics of Australian federal electorates. We have isolated four dimensions—ethnicity, socio-economic status, familism and urban-rural differences—representing the main headings under which the constituencies may be classified, and measured each with reliable multiple item scales.[17] The overall results are consonant with findings in other countries, although the reduction of the full range of social structural variables to only four is perhaps more parsimonous than usual. For example, in an analysis of the 624 British parliamentary constituencies, Webber and Craig (1976) identified different structural variables that led to a classification of 30 different constituency types, each with a different social structure, and a distinctive socio-economic environment. By comparison, the social structure of Australian federal electorates is relatively simple.

The delineation of these four dimensions has clear implications for other aspects of social science research. For researchers working with individual level survey data, information on these contextual characteristics can be readily incorporated in a matter of minutes, using simple and readily available techniques.[18] Once incorporated, contextual variables can then be treated in the same way as ordinary, individual level attributes and analyzed just like any other variables, for example, by simple cross-tabulations or ordinary least squares regression.[19] While more complex techniques have often been used in the analysis of contextual effects, it is now known that the simple procedures are in fact the technically correct ones as well (Igra, 1980).

Much more research needs to be carried out before the impact of contextual effects on the wide range of social and political behaviour can be adequately understood and measured. To date the burgeoning international literature on the topic, mainly centred on Britain and the United States, has not been extended to Australia. By delineating the major social structural dimensions in Australian constituencies, we hope that this paper will prove a small step in stimulating research on the influence of contextual effects in Australian society.

NOTES

1. Although previous research in Britain has often used ad hoc collections of census variables, there are great dangers in this atheoretical approach. First, with only a limited number of constituencies (124 in Australia) and highly correlated variables (correlations of .7 and .8 are common), the statistical problems of multi-collinearity are very great and it

easy to obtain unstable and misleading results (e.g. Hanushek and Jackson, 1977: 86-96, 131-133.). This is especialy true when, as is commonly the case, stepwise regression procedures are used (ibid.: 95-96). Second, it is easy to mistake the meaning of an isolated variable. For example, in studies of voting behaviour constituencies with a relatively high proportion of the population having no car tend to vote Labor while those with a relatively high proportion having three or more cars tend to vote Liberal. It is natural to interpret these as reflecting the influence of socio-economic status, no cars indicating poverty and three cars indicating affluence. Yet both these interpreta- tions are grossly misleading. As we will later show, no cars indicates low familism and three cars indicates rurality; neither is at al' closely tied to socio-economic status.

2. For example, in Australia, the proportion of a constituency employed in trades is very closely correlated with voting behaviour. It is tempting to treat that as showing the influence of low socio-economic status but other measures of socio-economic status have lower (but still substantial) correlations, one much like the other. So it seems that proportion in trades taps more than just socio-economic status, possibly union membership as well, and it would therefore be very misleading to use the proportion in trades as the sole socio-economic status indicator.

3. The often cited dangers of using aggregate level data to infer individual level effects (e.g. Robinson, 1950) are greatly exaggerated (Firebaugh, 1978; Hanushek and Jackson, 1977: 84-86). The essential point of these rather technical arguments is that metric regression coefficients (b's) and related metric statistics (e.g., percentage differences), are unaffected by aggregaion under most circumstances. Only standardized coefficients (betas, per cent of variance explained, etc.) are vulnerable to aggregation biases.

4. This will occur whenever there is a contextual effect and the contextual variable is correlated with an individual level variable of interest. In such cases, omitting the contexual variable is a specification error equivalent to omitting an individual level variable which has an effect on the dependent variable and is correlated with one or more of the independent variables (Igra, 1980). For example, if the socio-economic status of a constituency influences the voting behavior of people living in it, as much research in Britain and Australia seems to show, then it is misleading to estimate the effect of an individual's education, occupation, or

income on voting without taking this into account. The difficulty is that the occupational status of individuals is correlated with the socio-economic level of the constituency in which they live. Hence, estimates which ignore the socio-economic level of the constituency will wrongly attribute some of the constituency effect to the individuals's occupational status, which will then appear more influential than it actually is.

5. This, together with the detailed nature of the variables available, eliminates most of Bowman's (1969 objections, which pertain essentially to data base on small samples.

6. In ecological analysis there is a further question as to whether the results are sensitive to the particular choice of ecological unit. But in practice it seems that results based on small units (e.g. the equivalent of collectors districts) show essentially the same pattern as those based on broad units like parliamentary constituencies (Janson, 1980: 441-444; Sweetser, 1969). Furthermore, at least in respect to voting it seems that the contextual effect of constituencies rather than smaller units are in practice the crucial ones (Miller, 1978).

7. For a description and code of all the variables used in this paper, see (ABS, 1978: 24-32). All variables were converted to percentages of the total population in each constituency, except for those variables covering occupation (converted to percentages of the total labour force) and housing amenities (converted to percentages of total dwellings).

8. In general, we abandoned variables which were idiosyncratic in the sense of having no clear correlations with other variables and no obvious interpretation of their own, as well as those which were heterogeneous in the sense of being closely tied to two (or more) factors. This is the appropriate and conventional procedure in this context and, while a substantial element of judgment is involved, we believe we have not omitted any especially consequential variables. Further details are available on request.

9. Details of these analyses are available on request.

10. A very few groups, in all comprising less than four per cent of the overseas born population, are excluded. Some were so small that no reliable conclusions could be drawn even with census data and a few others (e.g. Americans) showed no distinctive residential patterns.

11. There are some further complexities. For example, there are grounds for suspecting that many of the Eastern European groups are disproportionately Jewish; that many were born in German displaced persons camps; and that the total number of Eastern Europeans may be somewhat underestimated, Balts and Ukranians being disinclined to acknowledge birth in the USSR and appearing instead as 'Other Europeans' (Jupp, 1966 and personal communication). Jupp also suggests that the similarities between Greeks, Egyptians, and Cypriots may come about because the latter two groups are largely of Greek ethnicity.

12. We also constructed factor scores, each based on a separate principal components analysis of the appropriate ethnic variables. These are very highly correlated with the simpler scales described in text (r=0.98 between the two Mediterranean scales, 0.99 between the Eastern European scales, and 0.99 between the Northern European scales) and in practice are virtually interchangeable. Details are available on request.

13. The reason for using factor scores, rather than simply adding the variables up as we did for the ethnicity scales, is that the component items have different metrics and, unlike the ethnic variables, their sum has no intuitively clear meaning. However, the factor scores are just a weighted sum of items and the weights are all nearly equal (Table 16.4), so in practice the two procedures are essentially identical. Constructing scales by some other reasonable procedure (weighting by the inverse of the items' standard deviations, for example) would also lead to essentially the same results in practice.

14. This linear transformation takes the form:

Transformed score = (raw score + 2.6393)/18.2436.

15. The procedures are idential to those used in constructing the socio-economic status scale and subject to the technical qualifications and comments mentioned above. The raw scale was transformed by a linear transformation taking the form:

Transformed score = (raw score + 14.1531)/24.6883

16. The transormation was:

Transformed score = (raw score - 4.9139)/37.8869

17. The only data heretofore available on parliamentary constituencies (Legislative Research Service, 1980) are simple listings, ranking them in order according to a few selected census variables. Our procedure has the great advantage of reducing the large number of census variables which might logically be used in such <u>ad hoc</u> listings to a small number of theoretically meaningful diemensions, each reliably measured.

18. The example, using SPSS, the most widely available statistical package, information on the socio-economic status of a respondent's constituency can be created quite simply using only two control cards. By way of illustration, suppose that the constituency in which the respondent lives is identified according to the ABS's identification number (first column in the Appendix) in a variable called LOCATION. Then a measure of the socio-economic context, call it SESCON (fifth column of numbers in the Appendix), is obtained simply by the two commands:

```
COMPUTE  SESCON = LOCATION

RECODE SESCON (1=37) (3=57) (4=21) ... (126=11)
```

Similarly, a measure of Mediterranean ethnicity MEDETH would be created by:

```
COMPUTE  MEDETH = LOCATION

RECODE MEDETH (1=5.5) (3=4.2) (4=0.6) ... (126=0.2)
```

19. For example, suppose one wants to estimate the effect of the constituency's socio-economic composition (SESCON) respondent's voting behaviour (VOTE) controlling for conventional individual level variables—e.g. respondent's own education (ED), occupational status (OCC), and ideology (IDEOL) — and also controlling for contextual characteristics other than socio-economic status—e.g. the proportion in the constituency of Mediterranean origin (MEDETH). Then the appropriate regression, again in SPSS, is simply:

```
REGRESSION VARIABLES = VOTE, SESCON, ED, OCC,
   IDEOL, MEDETH/REGRESSION = VOTE WITH SESCON,
   ED, OCC, IDOL, MEDETH(2)/
```

Tabular and other forms of analysis follow the same simple pattern: the contextual variables are treated in exactly the same way as ordinary individual level variables.

REFERENCES

Australian Bureau of Statistics **Characteristics of Population and Dwellings: Commonwealth Electoral Divisions** (Canberra: ABS), 1971.

Australian Bureau of Statistics **Collection District Summary Files Final Data Technical Details** (Canberra: ABS), 1978.

Barnett, M.J. "Aggregate Models of British Voting Behaviour", **Political Studies** 11:2, 1973, pp. 121-134.

Bohrnstedt, George W. and T. Michael Carter, "Robustness in Regression Analysis." pp. 118-146 in Herbert L. Costner (ed.), **Sociological Methodology 1971.** (San Francisco: Jossey-Bass), 1971.

Bowman, R.A., "Factoring the Factors: Factorial Ecologies and Ethnic Research". **Australian and New Zealand Journal of Sociology** 14.2, 1969, pp. 191-193.

Bowman, R.A. and P.L. Hosking "A Factorial Ecology of Auckland" in R.J. Johnston and J.M. Soons (eds.), **Proceedings of the Sixth New Zealand Geography Conference Vol. 1** (Christchurch: New Zealand Geographical Society), 1971, pp. 273-280.

Burnley, I.H., "Ethnic Settlement Formation in Wellington-Hutt". **New Zealand Geographer,** 28 (1972 October): 151-170.

Burnley, I.H., "The Social Ecology of Immigrant Settlement in Australian Cities' in I.H. Burnley (ed.), **Urbanisation in Austrialia** (London: Cambridge University Press), 1974.

Butler, D. and D.E. Stokes, **Political Change in Britain** (London: Macmillan), 1974.

Cox, K.R. "The Neighbourhood Effect in Urban Voting Response Surfaces" in D.C. Sweet (ed.), **Models of Urban Structure** (Lexington, Mass.: Heath), 1972.

Crewe, I. and C. Payne, "Analysing the Census Data" in D. Butler and M. Pinto-Duschinsky, **The British General Election of 1970** (London : Macmillan) 1971.

Firebaugh, G., "Structural Determinants of Urbanisation in Asia and Latin America, 1950-70"2 **American Sociological Review** 44:2, 1979 pp.199-215.

Fitton, M., "Neighbourhood and Voting: A Sociometric Explanation" **British Journal of Political Science** 3:4, 1973, pp. 445-72.

Frisbie, W.P. and L. Neidert, "Inequality and the Relative size of Minority Populations: A Comparative Analysis" **American Journal of Sociology** 82:5, 1977, pp. 1007-1030.

Hanushek, E.A. and J.E. Jackson, **Statistical Methods for Social Scientists** (New York: Academic Press), 1977.

Harman, H.H., **Modern Factor Analysis,** (Chicago: University of Chicago Press), 1976.

Igra, Amon, "Three methods of estimating a model of group

effects: A comparison with reference to school effect studies", **Sociological Methods and Research** 9:1, 1980, pp.61-86.

Janson, C.G. "Some Problems of Ecological Factor Analysis" in M. Dogan and S. Rokkan (eds.), **Quantitative Ecological Analysis in the Social Sciences** (Cambridge, Mass: MIT), 1969.

Janson, C.G. "Factorial Social Ecology: An Attempt at Summary and Evaluation", **Annual Review of Sociology** 6:4, 1980, pp.433-56.

Jones, F.L., "Ethnic Concentration and Assimilation: An Australian Case Study', **Social Forces** 45:3, 1967, pp.412-23.

Jones, F.L., **Dimensions of Urban Social Structure** (Canberra: ANU Press), 1969.

Jupp, James, **Arrivals and Departures** (Melbourne: Cheshire-Lansdwne), 1966.

Kim, J.O. and Charles W. Mueller, Factor Analysis: Statistical Methods and Practical Issues. (Beverly Hills: Sage University Papers No. 07-014), 1978.

Lawley, D.N. and A.E. Maxwell, **Factor Analysis as a Statistical Method**, (London: Butterworth), 1971.

Legislative Research Service, **Commonwealth Electoral Divisions: Comparisons of Selected Characteristics of the Population, 1976 Census**, (Canberra: Parliamentary Library Legislative Research Service Basic Paper No. 6), 1980.

Lincoln, J.R., "Community Structure and Industrial Conflict: An Analysis of Strike Activity in SMSAs" **American Sociological Review** 43:2, 1978, pp.199-220.

Linz, J., "Ecological Analysis and Survey Research' in M. Dogan and S. Rokkan (eds.), **Quantitative Ecological Analysis in the Social Sciences** (Cambridge, Mass:MIT), 1969.

Logan, M.I., C.A. Maher, J. Mckay, and J.S. Humphreys, **Urban and Regional Australia: Analysis and Policy Issues** (Malvern, Victoria: Sorrett), 1975.

Miller, W.L., **Electoral Dynamics in Britain** (London: Macmillan), 1978.

Nachmias, C., 'Community and Individual Ethnicity: The Structural Context of Economic Performance' **American Journal of Sociology** 85:3, 1979, pp.640-652.

Przeworski, A. and G.A.D. Soares, "Theories in Search of a Curve: A Contextual Interpretation of Left Vote' **American Political Science Review** 65:1, 1971, pp.51-68.

Putnam, R.D., "Political Attitudes and Local Community', **American Political Science Review** 15, 1966, pp.351-7.

Robinson, W.S., 'Ecological Correlations and the Behaviour of Individuals' **American Sociological Review** 15, 1950, pp.351-7.

Rasmussen, J., 'The Impact of Constituency Structural Characteristics upon Political Preferences in Britain' **Comparative Politics** 5:1, 1973, pp.123-45.

Segal, D.R. and M.W. Meyer, 'The Social Context of Political Partisanship' in M. Dogan and S. Rokkan (eds.), **Quantitative Ecological Analysis in the Social Sciences** (Cambridge, Mass.: MIT Press), 1969.

Selvin, H.C. and W.O. hagstron, 'The Empirical Classification of Formal Groups' **American Sociological Review** 28:3, 1963, pp.399-411.

Sewell, William H. and J. Michael Armer, "Neighbourhood Context and College Plans", **American Sociological Review**, 31:2, 1966, pp.159-168.

Shevsky, E. and W. Bell, **Social Area Analysis** (Stanford: Stanford University Press), 1955.

Sly, P.F. and J. Tayman, 'Ecological Approach to Migration Reexamined', **American Sociological Review**, 42:4, 1977, pp. 783-95.

Studlar, D.T., 'Social Context and Attitudes towards Coloured Immigrants', **British Journal of Sociology** 28:2, 1977 pp.168-184.

Sweetser, F.L., "Neighbouhood Typologies and Social Ecological Theory", Paper presented to the 9th World Congress of Sociology, Uppsala, Sweden, 1978.

Sweetser, F.L., 'Ecological factors in metropolitan zones and sectors' in M. Dogan and S. Rokkan (eds.), **Quantitative Ecological Analysis in the Social Sciences**, (Cambridge, Mass: MIT Press), 1969.

Taylor, P.J. and R.J. Johnston, **The Geography of Elections**, (London: Croom Helm).

Timms, D.W.G., **The Urban Mosaic** (Cambridge: Cambridge University Press), 1971.

Trlin, A.D., 'Dutch Immigrants in Auckland: A Factorial Ecology' **New Zealand Geographer** 31 (October), 1975, pp.124-141.

Webber, R. and F. Craig, A **Socio-economic Classification of British Parlimentary Constituencies.** (London: Office of Population and Census), 1976.

Weight, Gerald C., Jr., "Contextual Models of Electoral Behavior: The Southern Wallace Vote", **The American Political Science Review**, 71:2, 1977, pp.497-508.

APPENDIX
Scale Scores for all Federal Electorates

Census Code No.	Electorate	Birthplace Groups[a] Mediterranean	Eastern Europe	Northern Europe	Socio-economic status[b]	Familism[b]	Rural-Urban[b]
1	ADELAIDE SA	6.4	2.0	14.0	37	54	10
3	BALACLAVA VIC	4.8	5.2	9.5	57	44	16
4	BALLARAT VIC	0.9	0.5	5.1	21	51	33
5	BANKS NSW	3.9	0.7	7.1	17	70	8
6	BARKER SA	1.5	0.6	6.4	14	55	87
7	BARTON NSW	8.2	0.8	8.2	31	48	9
8	BASS TAS	0.8	0.7	7.5	21	61	21
9	BATMAN VIC	14.2	1.0	6.0	27	41	8
10	BENDIGO VIC	0.8	0.5	3.9	19	48	41
11	BENNELONG NSW	4.1	1.5	8.9	57	63	10
12	BEROWA NSW	2.0	1.2	10.1	70	79	15
13	BLAXLAND NSW	10.9	2.1	7.6	8	52	6
14	BONYTHON SA	2.6	0.8	38.1	18	93	12
15	BOOTHBY SA	2.9	1.3	10.7	70	67	20
16	BOWMAN QLD	1.2	0.7	10.1	22	62	18
17	BRADDON TAS	0.3	0.4	6.0	16	61	29
18	BRADFIELD NSW	1.2	1.6	9.1	100	88	18
19	BRISBANE QLD	2.0	0.9	7.4	28	49	10
20	BRUCE VIC	7.3	1.4	9.2	49	96	13
21	BURKE VIC	14.6	1.3	8.3	9	77	13
22	CALARE NSW	0.7	0.7	3.9	20	46	42
23	CANBERRA ACT	3.6	1.9	11.4	62	88	11
24	CANNING WA	1.7	0.6	24.8	12	67	68
25	CAPRICONIA QLD	0.5	0.3	4.5	13	52	35
26	CASEY VIC	2.6	1.0	14.6	42	91	22
27	CHIFLEY NSW	5.3	1.4	11.6	0	70	4
28	CHISHOLM VIC	4.6	1.4	8.8	54	72	13
29	COOK NSW	1.9	0.8	9.2	40	76	16
30	CORANGAMITE VIC	1.0	0.6	7.0	27	63	65
31	CORIO VIC	6.9	1.8	11.7	19	59	12
32	COWPER NSW	0.3	0.2	4.1	15	45	41
33	CUNNINGHAM NSW	11.4	1.5	14.0	19	54	7
34	CURTIN WA	3.4	1.4	13.5	70	58	19
36	DARLING DOWNS QLD	0.3	0.3	4.6	19	53	46
37	DAWSON QLD	2.0	0.4	4.6	13	46	56
38	DEAKIN VIC	3.5	1.3	12.4	44	90	11
39	DENISON TAS	2.0	1.7	8.4	46	52	12
40	DIAMOND VALLEY VIC	5.6	1.4	7.8	56	100	10
135	DUNDAS NSW	3.6	1.0	7.8	44	69	10
41	EDEN MONARO NSW	2.6	1.0	5.6	20	47	42
87	FADDEN QLD	1.0	0.7	13.4	11	82	25
43	FARRER NSW	0.9	0.6	4.4	21	54	52
44	FISHER QLD	0.5	0.5	9.0	14	63	45
45	FLINDERS VIC	1.3	0.7	16.2	36	73	22
46	FORREST WA	2.7	0.9	11.4	13	55	72
47	FRANKLIN TAS	0.6	0.6	7.0	21	76	19
48	FRASER ACT	4.6	2.0	11.0	58	84	39
49	FREMANTLE WA	6.5	0.6	15.2	28	60	18
50	GELLIBRAND VIC	19.0	2.0	7.9	4	31	3
51	GIPPSLAND VIC	1.8	0.6	6.7	20	57	62
52	GRAYNDLER NSW	21.7	1.5	6.5	11	22	2

53	GREY SA	2.5	1.1	12.8	16	41	49
54	GRIFFITH QLD	5.2	1.6	8.6	20	19	8
55	GWYDIR NSW	0.5	0.4	2.5	13	42	80
56	HAWKER SA	5.0	1.4	11.3	32	54	11
57	HENTY VIC	8.9	2.4	8.6	33	48	11
58	HERBERT QLD	2.9	0.5	5.6	18	51	29
59	HIGGINS VIC	5.5	4.4	8.2	78	37	14
60	HINDMARSH SA	11.7	1.9	9.0	23	55	11
61	HOLT VIC	5.9	1.3	16.6	13	81	19
62	HOTHAM VIC	11.0	1.7	10.6	22	80	12
63	HUGHES NSW	2.6	0.7	11.8	22	72	8
64	HUME NSW	0.5	0.3	2.7	12	44	86
65	HUNTER NSW	0.7	0.8	6.6	14	56	13
66	INDI VIC	2.9	1.0	5.0	19	51	62
67	ISAACS VIC	2.6	0.8	13.9	35	69	12
68	KALGOORLIE WA	2.2	1.2	10.7	18	35	38
69	KENNEDY QLD	0.7	0.6	4.5	12	41	60
70	KINGSFORD-SMITH NSW	15.5	1.4	7.6	9	27	3
71	KINGSTON SA	1.3	0.8	22.1	29	96	15
72	KOOYONG VIC	6.1	2.7	7.8	82	45	15
73	LALOR VIC	17.0	2.8	11.9	7	72	8
75	LA TROBE VIC	1.8	0.9	18.8	30	92	21
76	LEICHHARDT QLD	3.7	0.7	5.4	17	36	40
77	LILLEY QLD	1.6	0.6	7.8	21	48	12
78	LOWE NSW	14.1	2.2	6.8	34	34	8
79	LYNE NSW	0.4	0.4	4.8	13	47	37
80	MACARTHUR NSW	1.3	0.7	12.5	19	63	26
81	MACKELLAR NSW	2.6	1.0	13.5	45	74	15
83	MALLEE VIC	2.8	0.3	2.8	16	48	93
84	MARANOA QLD	1.0	0.3	2.5	9	94	100
85	MARIBYRONG VIC	12.5	1.3	8.2	24	71	11
86	MACMILLAN VIC	3.2	0.9	10.2	16	60	49
136	MCPHERSON QLD	0.7	0.9	10.2	24	50	29
137	MACQUARIE NSW	2.5	1.4	14.2	25	73	10
88	MELBOURNE VIC	21.7	1.4	5.5	37	0	2
89	MELBOURNE PORTS VIC	13.3	3.7	9.6	34	1	3
90	MITCHELL NSW	3.0	1.1	10.0	35	82	27
91	MOORE WA	2.6	0.9	18.4	25	74	64
92	MORETON QLD	1.7	1.0	18.1	32	69	15
93	MURRAY VIC	3.6	0.3	3.9	15	58	83
94	NEWCASTLE NSW	3.9	1.1	5.9	25	36	6
93	NEW ENGLAND NSW	0.5	0.5	3.2	27	51	53
96	NORTHERN TERRITORY	2.8	1.5	7.5	30	19	17
97	N SYDNEY NSW	4.2	2.5	11.2	73	30	7
98	OXLEY QLD	1.2	0.9	11.9	7	58	15
99	PARRAMATHA NSW	7.8	1.2	8.9	16	62	6
100	PATERSON NSW	0.5	0.5	3.5	11	43	58
101	PERTH WA	8.6	1.7	15.3	26	55	15
102	PETRIE QLD	0.8	0.5	9.2	23	75	15
103	PHILLIP NSW	5.7	4.2	11.0	41	19	5
104	PORT ADELAIDE SA	7.5	2.5	16.0	5	54	7
105	PROSPECT NSW	12.9	2.1	9.7	4	63	10
106	REID NSW	11.2	2.2	7.0	5	38	3
107	RICHMOND NSW	0.9	0.3	4.1	15	48	46
108	RIVERINA NSW	4.0	0.3	2.8	15	63	72
109	ROBERTSON NSW	0.6	0.4	9.2	17	49	16

110	RYAN QLD	1.1	1.3	9.5	74	71	16
111	SCULLIN VIC	17.6	0.7	6.1	11	77	10
112	SHORTLAND NSW	0.9	0.5	6.9	19	63	10
113	ST GEORGE NSW	14.3	1.0	6.9	19	44	7
114	STIRLING WA	5.8	0.7	18.6	23	83	16
115	STURT SA	9.2	1.7	12.5	45	73	17
116	SWAN WA	4.6	1.6	17.1	18	55	13
117	SYNEY NSW	16.0	1.4	6.3	24	1	0
118	TANGNEY WA	2.7	0.8	23.9	27	87	18
119	WAKEFIELD SA	2.0	0.5	6.3	14	50	95
120	WANNON VIC	0.4	0.3	3.7	17	53	76
121	WARRINGAH NSW	2.5	1.4	12.5	58	47	10
122	WENTWORTH NSW	5.4	5.4	11.0	62	4	8
123	WERRIWA NSW	4.7	1.2	13.2	11	65	8
124	WIDE BAY QLD	0.5	0.3	5.3	12	49	43
125	WILLS VIC	24.2	0.9	4.6	9	31	3
126	WILMOT TAS	0.3	0.4	4.8	11	49	52
	Mean	4.9	1.2	9.4	27.2	43.7	26.3
	Standard Deviation	5.2	0.9	5.1	18.9	20.2	24.0

Notes:

a Scored as the per cent of the constituency's population born in the indicated countries. See Table 16.2 for the countries and the text for details.

b Factor weighted scores; see Table 16.4 and the text for details. For ease of interpretation, scores are transformed into a conventional 0 to 100 range. Thus the constituency with the highest average level of education, occupation, and income gets a score of 100 on the socio-economic status scale and that with the lowest average gets 0. The constituency with the largest proportion of family type suburban dwellings gets a score of 100 on the familism scale, while the least familistic gets 0. Finally, the constituency with the largest proportion of farmers, fishermen and the self-employed gets 100 on the rural-urban scale, while that with the fewest gets 0.

Source: As for table 1.

Index